Getting Started with the MSP430 Launchpad

Getting Started with the MSP430 Launchpad

Adrian Fernandez
Dung Dang

AMSTERDAM • BOSTON • HEIDELBERG • LONDON
NEW YORK • OXFORD • PARIS • SAN DIEGO
SAN FRANCISCO • SINGAPORE • SYDNEY • TOKYO
Newnes is an imprint of Elsevier

Newnes is an imprint of Elsevier
The Boulevard, Langford Lane, Kidlington, Oxford, OX5 1GB, UK
225 Wyman Street, Waltham, MA 02451, USA

First published 2013

Notices
Knowledge and best practice in this field are constantly changing. As new research and
experience broaden our understanding, changes in research methods, professional practices,
or medical treatment may become necessary.

Practitioners and researchers must always rely on their own experience and knowledge
in evaluating and using any information, methods, compounds, or experiments described herein.
In using such information or methods they should be mindful of their own safety and the safety
of others, including parties for whom they have a professional responsibility.

To the fullest extent of the law, neither the Publisher nor the authors, contributors, or editors,
assume any liability for any injury and/or damage to persons or property as a matter of products
liability, negligence or otherwise, or from any use or operation of any methods, products,
instructions, or ideas contained in the material herein.

British Library Cataloguing-in-Publication Data
A catalogue record for this book is available from the British Library

Library of Congress Cataloging-in-Publication Data
A catalog record for this book is available from the Library of Congress

ISBN: 978-0-12-411588-0

For information on all Newnes Press
visit our website at store.elsevier.com

This book has been manufactured using Print On Demand technology. Each copy is produced
to order and is limited to black ink. The online version of this book will show color figures
where appropriate.

Working together
to grow libraries in
developing countries

www.elsevier.com • www.bookaid.org

CONTENTS

In 2010, yours truly started on a new and exciting project at Texas Instruments. The objective was clear: we needed to create a new affordable and easy-to-use microcontroller development platform that can be accessible by thousands of engineers. This is how the MSP430 LaunchPad Evaluation Kit came about. We knew its affordability as well as user-friendliness was very attractive, but the results surpassed even our greatest expectations! After 2 years, our LaunchPads are in the hands of more than a quarter million engineers and students.

Immediately after the release, we realized that the LaunchPad found itself in the middle of a microcontroller revolution, where microcontroller development is becoming more accessible, affordable, and simple, enabling not only embedded design experts but also enthusiasts from all disciplines and levels of background to actively participate.

The success of the LaunchPad did not simply stop there with the hardware story. In order to supplement the professional lines of industrial-grade software development tools by Texas Instruments and other vendors, the open-source community birthed Energia, which made another huge stride in bringing microcontroller development even closer to the masses. Energia makes programming a microcontroller a nonintimidating task for nonengineers, college and high-school students, and in some instances even 7-year-old kids.

Another contributing factor to the LaunchPad's success, and we think this might be the most important one, is the extremely friendly and supportive community that surrounds the LaunchPad. The LaunchPad can only be as successful as its users, and the ever-growing community contributing so much content, tools and projects is the best indication of how the LaunchPad has been welcomed into the desks of microcontroller developers around the world.

Having the unique opportunity to start and continue to contribute to the success story of LaunchPad, we, the authors, feel that it's our duty to help bring the LaunchPad to even more interested users, regardless of their background or expertise. To continue and foster this growth, we'd like to help everyone get familiar with the LaunchPad, learn all of the things the LaunchPad is capable of, and understand how to apply such capabilities to create things that they have always wanted to make. And we wanted to capture all of this into something that is simple to follow, easily digestible and hopefully a tiny bit inspiring. This is how this book came about.

Happy making!

Adrian Fernandez
Dung Dang
Dallas, TX
March 2013

ACKNOWLEDGMENTS

This book would not have been possible without many key individuals and groups.

First, special thanks to Robert Wessels for creating the Energia software development environment for the LaunchPad Evaluation Platform. Energia has opened the doors to microcontroller developers of all experience levels, thanks to its intuitive interface and familiar functions and APIs. Additionally, special thanks to all contributors to Energia and the greater open-source community, who have created a strong foundation in building out this framework and the functional APIs that have enabled more people to start tinkering with microcontrollers.

In addition, thanks to Texas Instruments for creating the LaunchPad Evaluation Platform, which has provided affordable, yet powerful, microcontroller development kits. By providing these tools at accessible pricepoints, more individuals are able to get their hands dirty with microcontroller development!

Thanks to Circuitco Electronics for creating the Educational BoosterPack. This plug-in module for the LaunchPad has further enhanced what is possible with the LaunchPad by providing so great many features in an affordable package.

Special thanks to @bluehash and the 43oh.com online community for being an active, friendly and enthusiastic community. This community has helped thousands of new and experienced microcontroller developers alike.

And of course, thanks to our family and friends who supported us during the writing of this book. Thanks to Ellen and Kacy for putting up with our late nights of tapping away on our keyboards and everything they have done for us! And thanks to Ollie and Buck, the two best dogs any author could ask for!

Thank you!

Adrian Fernandez
Dung Dang

CHAPTER 1

Prepare for Liftoff!

However you came here, we're glad that you did! In the following pages, we're going to learn a lot about a little red board that we call the MSP430 LaunchPad Evaluation kit. For the purpose of this book, we'll call our new friend "LaunchPad".

We've got a long way to go, but if you stick with us, we think that you and the LaunchPad can become the best of friends. This could be the beginning of a special bond filled with flashing LEDs, buzzing buzzers, and spinning motors!

May you embark on a journey full of discovery and wonder as you conquer the ultimate quest alongside your trusty LaunchPad.

1.1 WHO SHOULD READ THIS BOOK?

The LaunchPad is a great introduction to the world of microcontrollers. Interested readers from all experience levels may find this book a great resource to help guide them through their journey with the LaunchPad.

You came to the right place if:

- You are a student or a curious reader without any background or experience with electronics or microcontrollers.
- You are a hobbyist who wants to get their hands dirty making new electronics by yourself. This book can help you get started.
- You are an electrical engineer who has focused on circuits and hardware but have not yet had a chance to step into the world of embedded programming. This book can reveal the simple steps needed to get started with coding and giving your design some intelligence.
- You are a software developer who can write code in your sleep but have not had a chance to touch the hardware side of things. This book can show you how easy it is to make the first step over to the hardware side.

- You come from a different professional discipline and want to incorporate electronics into whatever it is that you do. This book can be a good way for you to learn about electronics and may spark some creative ideas on how to incorporate this new-found knowledge into unique avenues.

This book is suitable for a wide variety of audiences, regardless of age, background, or level of knowledge on electronics or microcontrollers. As long as you are interested in learning, we think that this book could be a great tool for you.

1.2 WHAT TO EXPECT OUT OF THIS BOOK

This book will show you how to have fun with your LaunchPad and help you learn a thing or two along the way. By the end of this book, you will be able to do the following:

1. Know what a LaunchPad is and what it is capable of. Most importantly, we want you to know what you can make with the LaunchPad.
2. Understand what it takes to create cool things with your LaunchPad. We'll learn how to teach our LaunchPad new tricks, explore the inner workings of the LaunchPad, and control how the LaunchPad behaves and interacts with the outside world.
3. Make simple but fun real-world applications with the LaunchPad. With our LaunchPad and clever coding, we can solve lots of problems. Are you down for a challenge?

> **Challenge:**
> Think of one thing that you really, really want to make. Now write it down! What is the main purpose of your dream "thing"? What does it do? How does it behave or interact in different scenarios? Using only what you know as of now, describe how you would implement it, what components you think it should have, and how they work together.
> My dream creation: _____

By the end of the book, we will come back to this idea. Except by the end of the book, we will redo the exercise using all of the new concepts that you have learned. We will compare the notes and see if there's any difference in the way you approach the problem.

Ultimately, we hope that with your newly gained microcontroller knowledge, and more importantly, a newly gained perspective in electronics and problem solving, you will be able to clearly identify the pieces of the puzzles and become much more equipped to go off and make the next great microcontroller application.

1.3 WHAT IS POSSIBLE?

Look around you, electronics are everywhere and help make the world go round. From the alarm clock we press 'snooze' on every morning to the electric toothbrush that we use before we go to bed. Many of these electronics exist because of microcontrollers. And as the name implies, microcontrollers are little things that control something.

Our LaunchPad is a microcontroller development kit that allows users just like you to start exploring the world of microcontrollers. Every day we interact with hundreds of microcontroller-enabled products. Sometimes microcontrollers are hidden away, behind the scenes. Maybe they're behind your steering wheel getting ready to deploy an airbag right when you need it. Or maybe it's in the smoke detector that ensures the safety of you and your family. These versatile devices are going to be at the center of the projects that we will develop throughout this book.

Let this book be the first step toward your microcontroller dreams. This book offers a first look at the MSP430 LaunchPad evaluation kit from Texas Instruments. The LaunchPad was the outcome of TI engineers with years of microcontroller and embedded systems experience, with hopes of providing an easy-to-use, fun, and affordable way to start making things with electronics.

LaunchPad is part of a revolution where microcontroller development is becoming easier, enabling anyone who's interested to become a microcontroller expert. If this book does one thing right, let it be that this book convinces you that you can do it!

CHAPTER 2

Meet the LaunchPad

Let's take some time to meet the LaunchPad. As we hinted at in the last chapter, we think that the LaunchPad will make a great companion as you start your journey to learn more about microcontrollers and electronics.

In a way, the LaunchPad is essentially a complete system that acts and behaves very similarly to a human. The most interesting parts of our LaunchPad include the brain, the senses, and the actions.

It interacts with the world through its integrated senses, processes the received information, makes decisions, and ultimately takes actions based on those decisions.

2.1 THE BRAIN, THE SENSES, AND THE ACTIONS

2.1.1 The Brain

At the center of the LaunchPad is the MSP430 microcontroller. We can think of the microcontroller as the brain of our LaunchPad kit. The brain is responsible for processing, contemplating, and making all decisions for the LaunchPad based on the information received from the available senses. Once the brain is done processing the information, it is capable of making decisions and performing various actions.

The specific MSP430 chip that makes up the brain of our LaunchPad is the MSP430G2553. We'll get more acquainted with its unique features and capabilities in the next chapters.

2.1.2 The Senses

Our LaunchPad comes with a couple of built-in senses. These senses provide information to our LaunchPad brain to help describe the world around it. The LaunchPad can feel touch, sense temperature, or hear communications from other systems. The LaunchPad can feel touch thanks to the two push buttons found at the bottom of the board. Temperature can be sensed by the MSP430 microcontroller chip itself because of a temperature sensor available inside. And lastly, the LaunchPad can hear commands from external systems when connected to a computer with the provided USB cable, among other things.

In addition to the out-of-box capabilities of your LaunchPad, you can enhance your LaunchPad with new senses by equipping it with new modules thanks to two rows of breakout headers, or connectors, found on each side of our LaunchPad. For example, some additional senses could be hearing sound, detecting motion, or seeing light. These new senses can make your system more aware of the world around it, by giving the LaunchPad brain more information to process. This additional data allows our LaunchPad to make more accurate decisions and carry out the appropriate actions. These add-on modules are called "BoosterPacks," which follow a standardized footprint that make it easy to plug in new capabilities to the LaunchPad.

2.1.3 The Actions

When the brain is done processing the sensed information, it usually needs to make a decision whether to act on it. In most occasions, the brain needs to somehow let the user, the world, or other external systems know that something has happened or is happening. For this type of notification, the LaunchPad relies on its available actions to perform such tasks. The actions included with the LaunchPad are the two LEDs that can provide visual effects, and the verbal commands that the LaunchPad can send back to external systems through the USB cable or other interfaces.

Similarly to the senses, you can also equip your LaunchPad with more actions to make your application more interactive and useful for your needs thanks to the modularity of BoosterPack plug-in modules. The same two rows of breakout headers let you connect external modules or electrical components that can carry out more enhanced actions such as motor control, make sounds on a speaker, or even wireless communication to other LaunchPads or systems.

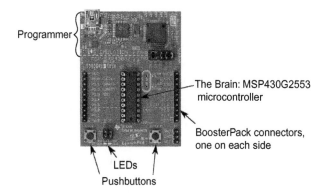

Programmer

The Brain: MSP430G2553 microcontroller

BoosterPack connectors, one on each side

LEDs

Pushbuttons

2.2 TEACH THE LAUNCHPAD

For our LaunchPad to do anything meaningful, it needs to take in all types of data from various senses and understand what to do with this collection of information. If the LaunchPad senses that temperature is getting too hot, should it turn on a light, spin a fan motor, or sound an alarm? Maybe all of the above.... We can teach our LaunchPad to act appropriately based on the information coming in from the many senses by programming it.

To teach our LaunchPad brain, a **programmer** is provided. The programmer is everything above the white dotted line on our LaunchPad kit, and it allows us to program our LaunchPad brain using the included USB cable. With this ability to teach and train your LaunchPad through the provided programmer, you have the power to create all types of applications! Looking around us, many of the products we use everyday are made of the same basic components—speakers, LEDs, motors, etc. However, it's the clever programming that allows us to truly innovate!

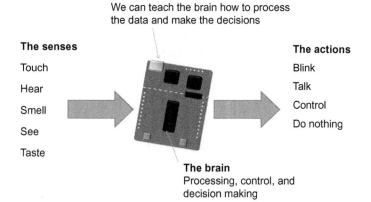

Teach it!
We can teach the brain how to process the data and make the decisions

The senses

Touch

Hear

Smell

See

Taste

The actions

Blink

Talk

Control

Do nothing

The brain
Processing, control, and decision making

2.3 INSIDE THE BOX

Your LaunchPad is a remarkable tool, it is capable of doing numerous activities. To accomplish some of these tasks, the LaunchPad is packaged alongside additional items inside of the box:

- USB Cable: The link to connect your LaunchPad brain with a computer. The link allows you to teach and program your LaunchPad. It also creates a communication channel that lets the LaunchPad and the computer exchange messages and commands that might be helpful during development.
- An extra MSP430 chip, MSP430G2542: Consider this as an extra brain that can be swapped out with the existing brain on the LaunchPad. The two chips have slightly different capabilities, and each brain has its own memory, so you can essentially create two different applications out of your one LaunchPad kit.

Tip

A dip package removal tool might be a good investment to safely and efficiently remove/replace your MSP430 chips from the LaunchPad socket without damaging/losing any of the pins. An MSP430 with all of its pins intact is a happy MSP430.

- Male and female pin headers: Very useful if you want to connect your LaunchPad to the external components (i.e. to add new senses or actions). You can solder these pin headers onto the breakout headers on the LaunchPad. The male and female connections will let you easily connect external modules onto the LaunchPad this way, making it a simple plug-n-play exercise.
- A 32-kHz crystal: A highly accurate crystal that your LaunchPad can use to keep track of time. This will become handy when you need to keep an accurate clock. This is optional as the LaunchPad brain already has a built in clock; however, this external crystal can help provide further accuracy.

2.4 PROJECT 1: THE FIRST ENCOUNTER OF THE LAUNCHPAD KIND

With the included USB cable, let's go ahead and plug in our LaunchPad to a USB power source!

Now that we've powered up our LaunchPad for the first time, we can see that the LaunchPad has already been programmed with a certain mission. This demo code was written to give first-time LaunchPad users a glimpse of what is possible with the available senses and actions on the LaunchPad. For this project and this exercise only, you get to assume the role of the user, for you will face the rest of the projects head-on as the developer. So sit back and enjoy!

Tip

If you are not using a brand new LaunchPad, there is a possibility that the default program has been erased or replaced by something else. If you would like to restore the original program, go to http://booksite.elsevier.com/9780124115880.

2.4.1 First Signs of Life

The first thing to happen is that the LaunchPad's onboard red and green LEDs will blink alternatively signaling the board's heartbeats. At this point, the LaunchPad is waiting for interaction from you, the user! To introduce yourself to your new friend, simply press the pushbutton on the bottom-left corner of the LaunchPad—it should be labelled "P1.3" above and "S2" below the pushbutton. At this point, the temperature sensor found inside of the MSP430 microcontroller will read the temperature and remember that reading. Now, any changes from that starting temperature will be reflected by the green or red LED's brightness. The green LED glows more brightly if the temperature gets colder than the starting temperature. Alternatively, the red LED glows more brightly if it gets hotter. Breathing and heartbeats, we have confirmation that our LaunchPad is alive and kicking!

2.4.2 Getting Intimate with Your LaunchPad

To heat things up, simply rub the MSP430 chip on your LaunchPad. This should increase the temperature read by the temperature sensor inside. As the temperature reading increases, the red LED should start to glow. To lower the temperature, try using a compressed air can to cool down the MSP430. This should cause the green LED to glow more brightly.

We can also recalibrate our system by pressing the pushbutton S2 again. This will cause the brain to forget the previous "start temperature" and will record a new temperature. Now, all future temperature readings will be compared against this new baseline temperature. To see a video of this demo, visit: http://booksite.elsevier.com/9780124115880.

2.4.3 Flow Chart for the Demo

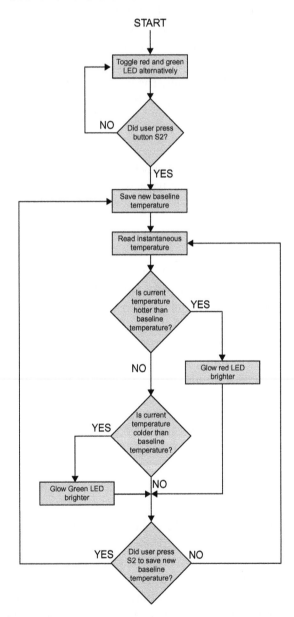

2.4.4 Relating the Demo Back to Our Senses and Actions

This sample program demonstrated several of the available senses and actions on our LaunchPad. Let's break it down!

Senses:
- Feel touch: Pressing the pushbutton to start the application and recalibrate the temperature.
- Sense temperature: This system is periodically checking the temperature reading to see if it is getting hotter or colder compared to the baseline temperature.

Actions:
- Blink: Depending on the changes in temperature, our LaunchPad's brain gives the command to turn on the red or green LED.
- [Optional] Talk to the PC: Even though we don't need to touch this here, the LaunchPad actually also reports the temperature back to the PC through the USB cable. We will learn how to teach our LaunchPad to talk to the computer in future chapters.

At this point, you have learned that our LaunchPad has a few senses and actions at its disposal. With some intelligence and cleverness, we can teach our LaunchPad to perform various actions based on the information gathered through its senses. We can simplify all of this activity into inputs and outputs or I/O. To help understand this concept, we can relate our LaunchPad's senses as "inputs" while our LaunchPad's actions are "outputs."

Now that we have a good idea of what our little LaunchPad is capable of, using its available inputs and outputs, we'll take some time to meet a few additional tools that will aid us through the rest of the book. In the next chapter, we will meet our first BoosterPack.

CHAPTER *3*

The Fellowship of the LaunchPad

Before we fully embark on our journey, let's take some time to meet a few of the tools that will prove to be very useful in our LaunchPad journey.

3.1 YOUR JOURNEY'S COMPANION

In this book, you have two options. You can choose to use a unique development tool that was created specifically for this book, or you can decide to make the many projects from the ground up. No matter which option you choose, you will have blinking LEDs and buzzing buzzers in no time!

3.2 THE EDUCATIONAL BOOSTERPACK

So far, we've learned that our LaunchPad has quite a few integrated actions and senses available right out of the box. However, throughout this book we will need the help of external components to introduce new functionality to our LaunchPad to help us solve more problems. We can do this through plug-in modules called BoosterPacks. BoosterPacks are modules that we can connect to our LaunchPad to add new capabilities such as displays, sensors, and more.

For the journey ahead of us, we'll be using the Educational BoosterPack.

The Educational BoosterPack will prove to be a very capable companion to our LaunchPad. It adds new inputs and outputs to our LaunchPad including:

- 16 × 2 character liquid crystal display (LCD)
- Piezo buzzer
- Potentiometer scroll wheel
- Microphone
- Red/green/blue light emitting diode (RGB LED)
- Three-axis accelerometer
- Alligator-clip-friendly input/output and ground pins.

We may not know what each of these items are or do quite yet, but we will have an opportunity to play with each of the Educational BoosterPack's features by the end of this book!

If you do not have an Educational BoosterPack, but would like to get one, you can find the BoosterPack here: www.boosterpackdepot.com.

The website above also hosts many resources pertaining to the BoosterPack including documentation and code examples. Also, the Appendix serves as a *Quick Reference* section, which has a dedicated section for the Educational BoosterPack. You'll find useful pin-mapping information and other things to help you get further acquainted with the Educational BoosterPack. Feel free to refer to that section as-needed throughout the course of this book.

For those that do not have an Educational BoosterPack, do not fret! There are other options as shown below.

3.3 THE BREADBOARD

A breadboard (sometimes called protoboard) is essentially the foundation to construct and prototype electronics. A breadboard allows for easy and quick creation of temporary electronic circuits or to carry out experiments with circuit design. Breadboards enable developers to easily connect components or wires thanks to the rows and columns of internally connected spring clips underneath the perforated plastic enclosure. The grid is made up of perfectly aligned spring clip holes that are 0.1″ apart in both the X and Y dimensions.

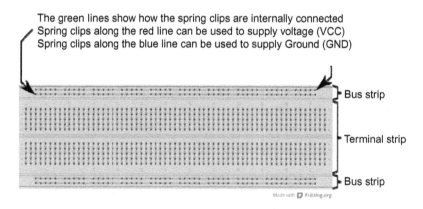

The green lines show how the spring clips are internally connected
Spring clips along the red line can be used to supply voltage (VCC)
Spring clips along the blue line can be used to supply Ground (GND)

Bus strip

Terminal strip

Bus strip

Made with Fritzing.org

A breadboard can be divided into two segments, which are called the *bus strip* and the *terminal strip*.

The *bus strip* consists of two long lines of spring clips running across the board, these lines can be used to provide supply voltage (VCC) and ground (GND) to the circuit. Typically, the supply voltage line is marked in red and the ground line is marked in blue (or black on some boards). All spring clips along each line are internally connected, making both supply voltage and ground signals conveniently accessible from any part of the breadboard.

What Is VCC?

VCC is the supply voltage that the microcontroller system uses. The LaunchPad can run properly when a supply voltage of 1.8–3.6 V is available. In most cases, the VCC of our MSP430-based projects will be ~3 V. We will use this terminology and referencing this supply voltage throughout the book.

What Is Ground?

All electrical current in a system needs to flow back to ground. In a typical microcontroller project, all components flow back to Ground, which is why this is usually referred to as "common ground." At GND, this is usually a 0-V reading. We will touch on VCC and GND throughout the book, so don't worry if this sounds foreign!

The *terminal strip* is the main area that can be used to populate the various circuit components. It is usually separated into two sides by a notch that runs along the middle of the board. Each side has many lines that are made up of five internally connected spring clip holes. The five spring clips on each line of the terminal strip are connected internally, allowing for component connections.

3.3.1 Connecting to a Breadboard

There are many breadboard-friendly components that are available for us to use and prototype with. We can easily identify breadboard-friendly components by their footprint.

Some through-hole-friendly components have long and wiry legs such as LEDs or resistors. Many common discrete components are typically very breadboard-friendly thanks to their bendable legs. You can easily bend the legs into the desired shape and size and insert them into any of the spring clip holes on the breadboard to enable different types of circuit designs.

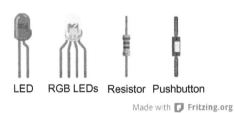

LED   RGB LEDs   Resistor   Pushbutton

Made with 🔲 Fritzing.org

In addition to some of the example components above, some integrated circuits (ICs) or chips also come in breadboard-friendly form-factors. The MSP430G2553 microcontroller found on our LaunchPad is one of those devices. The MSP430G2553 has legs that are perfectly spaced to fit right into the two rows of spring clips along each side of the notch that divides our terminal strip. The official package name for this type of chip is Dual-Inline Package (DIP). The spacing between the two rows of pins on a DIP package is 0.3″, which is exactly three times the spacing between the spring clips on the breadboard. This means that if needed, we can remove the MSP430G2553 device from our LaunchPad and plug it into a breadboard for further flexibility.

3.4 BREAKOUT BOARD

In reality, some components unfortunately do not come in the through-hole or breadboard-friendly DIP-style form factor. The reason is that many of the electronic components are ultimately made to be used on a printed circuit board (PCB) where components are ideally as small as possible and can be mounted directly on the surface of the board to minimize the footprint and conserve board space. These are called surface-mount components. While DIP devices and through-hole components are great for quick prototyping on a breadboard, they are not space efficient when it comes to creating the next slim and sexy electronic device that everyone wants for Christmas. As wonderful as it is to have breadboard-friendly components, many components are usually created to have the smallest physical footprint as possible. This smallness, however, comes at a price because they are not very good for quick prototyping and throwing one on a breadboard to carry out simple tests or experiments is a little bit harder.

To remedy this and enable easy prototyping for popular non-breadboard-friendly components, some manufacturers create breakout boards (which are in fact PCB themselves) that bring the pins of the component out to a 0.1″-spacing through-hole arrangement more suitable for breadboarding. A few example of breakout boards for common sensors are shown in the figure below. For those who are not using the Educational BoosterPack, we will actually be using the ADXL335 breakout board shown below for future projects that require an accelerometer.

Ambient light

Gyroscope

Accelerometer

Humidity temperature

Made with **Fritzing.org**

With the breakout boards, the dimensions and spacings now line up with the breadboard, but they're still just through holes. In order to enable connections whether directly to the breadboard, or using wires and/or clips, you still need to attach pin headers to the available through holes. We can add headers to the breakout board through a process called soldering.

3.5 SOLDERING

Soldering, in short, is the process of joining two metal components together by flowing a filler metal into the joint. This can be done with the help of a soldering iron. This is a hand-held tool that has a heating element at the tip. We can use this to melt a filler metal called solder into a joint that connects two metal components. This is possible because solder has a lower melting temperature than the metal components. In the electronics world, soldering connects the components not only mechanically but also electrically. This is ensured due to the fact that both sides of the components as well as the solder material are conductive and let electric current to flow through freely.

In circuit prototyping endeavors, we will encounter two common types of soldering: through-hole and surface mount soldering. Through-hole soldering is the easiest type, where components usually have long and accessible legs that can be easily placed into the through-holes. You will simply need to put the components in place and essentially fill the connections with solder. The latter type, surface mount, can become tricky when components get smaller and so do their pins or connection pads. This will require some practising and patience, so be sure to read up on some tutorials and check out some online videos to hone your soldering skills.

If you are new to soldering, we recommend starting off with through-hole soldering. Soldering headers onto breakout boards, which convert those tricky surface mount connections into breadboard-friendly through-holes, is a great way to build up your confidence with soldering and prototyping.

3.6 ELECTRICAL ETIQUETTE

When creating circuits, we have to be aware of the electrical etiquette that is defined in a document called a datasheet. All electronic components have a datasheet, which is a very detailed write-up that explains the specifications, characteristics, and proper usage of the specific component.

This documentation helps us understand several things. The first lesson is that not all pins are made equal. This especially applies to the legs of discrete through-hole components such as LEDs, diodes, capacitors, or transistors. Always make sure to figure out the positive (anode) and the negative (cathode) leads of the components. Inserting components with reversed polarity can cause a short circuit or excessive voltage applied to the components potentially resulting in failure, smoke, fire, and explosions! So be sure to always have your safety goggles on as precaution and, more importantly, always read the manual and device datasheets to ensure everything you build is right the first time.

A datasheet is the most important document for a device, it describes the device's complete mechanical specification, from dimensions, pinouts, functionality, block diagrams, operating conditions (supply voltage and temperature ranges) to electrical parameters and performance (power consumption, accuracy, speed, etc.). The datasheet also offers some basic circuit configurations to help you figure out how to connect and use the device properly. As it contains so much scientific information, reading a device datasheet won't be an easy and straightforward task to begin with. But if you start off with simple components, and with a little bit of patience, you will eventually be able to digest the datasheet and seek out the information that you need in order to use the device properly.

3.7 RECREATING THE EDUCATIONAL BOOSTERPACK

This book was written to heavily leverage the components on the Educational BoosterPack to illustrate different microcontroller and electrical concepts through exercises and project. On the other hand, we also wanted to emphasize the importance of you understanding the process of coming up with such circuits and ultimately be able to construct similar or new circuits when you start exploring and want to prototype your own unique design. For those without the Educational BoosterPack, we will also show you how to build the equivalent breadboard circuit for each project. So whether or not you have the BoosterPack, we highly encourage you to take a closer look at the breadboard circuit; and if you're daring, why not take a stab at rebuilding it from scratch?

3.8 SO I'VE GOT THE HARDWARE, WHAT ABOUT THE SOFTWARE?

Whether you plan on using the Educational BoosterPack or wiring this up on your own using a breadboard, we will be accomplishing a lot of things in the next few chapters. Now that you know what we will be dealing with in terms of hardware, what about the software side of things? Good question—in the next chapter, we will introduce you to Energia, which is the code editing software that we will use to write up code to teach our LaunchPad new tricks.

CHAPTER 4

Meet Energia—a Software Development Environment

4.1 HOW TO TEACH YOUR LAUNCHPAD NEW TRICKS

Now that we've been formally introduced to the LaunchPad and some of its companions, let's spend some time learning about how we can teach the LaunchPad's some new capabilities through software. While there are many tools that we can use to teach our LaunchPad new tricks, in this book we will be focusing on a simple and friendly development environment called Energia. With Energia, we can tell our LaunchPad which senses to use, tell it what to do with the information it has gathered, and finally teach it how to react to those senses.

Energia is an Integrated Development Environment (IDE) for the LaunchPad, which provides a clean and simple interface. This helps users get something up and running on their LaunchPad as quickly as possible without having to dig deep down into the pages of the microcontroller encyclopedia.

Energia is a community-developed and community-supported development environment that was created for the LaunchPad ecosystem. It is an open-source environment that is continuously being updated and improved upon, thanks to the collaborative efforts of individuals all over the world. As an open-source project, Energia has a very friendly license, which means that anyone can see the source code as well as have the opportunity to change or modify the project itself.

Without further ado, let's dive right in and get Energia running on your system.

4.2 GET ENERGIA

You can get Energia software for your computer at www.energia.nu. The website offers a few links that point to useful resources that can help you get more information on Energia and get started on the LaunchPad. You are encouraged to click around and see what information is available.

For the time being, our immediate interest is to download the software, which can be located at the *Download* link. As Energia continues to be enhanced and updated, make sure to get the latest version of the software on this download page. As you can see, there are three versions of Energia depending on the environment that you are using Windows, Linux, or Mac. Note that the software version as well as its release date correspond to the web page at the time this book was written. Download the version that corresponds to your particular setup.

HOME   GETTING STARTED   GETTING HELP   DOWNLOAD   PROJECT USING ENERGIA

energia-0101E0009-macosx.dmg - Mac OS X: Binary release version 0101E0009 (12/06/2012)

energia-0101E0009-windows.zip - Windows: Binary release version 0101E0009 (12/06/2012)

energia-0101E0009-linux.tgz - Linux: Binary release version 0101E0009 (12/06/2012)

4.2.1 Installation

Once you have downloaded the appropriate version of Energia, mosey on over to the "Getting Started" link in the navigation bar available at www.energia.nu.

This should point you to a wiki that explains the process for downloading and installing the appropriate drivers on your computer. Follow these steps available on the website.

Once you've installed the right drivers and downloaded Energia on your computer, unpackage/unzip the Energia download into a location that you can find later. One thing to note with Energia is that there is no install process, simply unzip and run the Energia executable.

4.3 PROJECT 2: LAUNCHPAD'S FIRST TRICK—BLINK

Now that you have Energia up and running on your computer, it's time to have some fun with the LaunchPad! For starters, let's teach our LaunchPad its first trick—blinking one of the onboard LEDs—in this case, the red one! This is a typical starter project for most new microcontroller engineers and is commonly referred to as the "hello world" of microcontroller development. Let's get going!

> **Note**
>
> "Hello world," a program that prints or displays "Hello world," is considered one of the simplest program in most (computer) programming languages. It has become a tradition to introduce beginners to the basic syntax of programming or to verify that the system is operational. In similar fashion, in the microcontroller world, blinking the LED has become the tradition for all beginners and professionals alike when bringing up a new system, making sure that the system is alive and kicking.

Ingredients:

1. LaunchPad

First, let's make sure that your LaunchPad is plugged in to your computer via USB and that Energia is looking in the right place so that it can start talking to your LaunchPad. We can do this by ensuring the following selection is made within the Energia menus:

Tools > Board > LaunchPad w/MSP430G2553 (16 MHz)

Now that Energia is set up properly, we can import one of the many prewritten code examples, or in Energia's term, sketches, that are provided inside of Energia. To import our blink LED example, we can follow the path shown below:

File > Examples > 1. Basics > Blink

> **Tip**
>
> Note that there are many preloaded example projects inside of Energia that span multiple categories! Feel free to import these various projects and teach your LaunchPad more tricks! These code examples are

available to help developers get started more quickly. They introduce new concepts and can be used as a starting point for your own development, rather than starting from scratch.

A sketch is what we call the code that we write inside of Energia and ultimately load into our LaunchPad brain. By selecting one of the preloaded sketches, Energia opens up a new window, which contains the *Blink* code example. Before we take a closer look at the code itself, let's go load this code into our LaunchPad brain to teach it its first lesson.

To get our code from Energia to the LaunchPad brain, a few things have to happen. First, Energia needs to convert the sketch into something that the LaunchPad brain can recognize. This process is called *compiling*. We can tell Energia to compile our code by pressing the "checkmark button" as shown in the screenshot below.

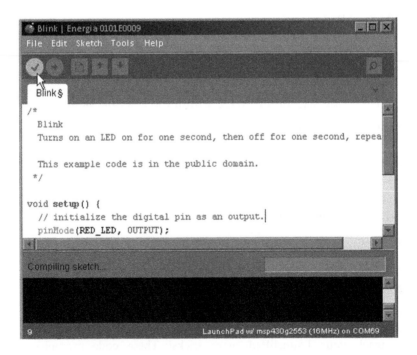

By compiling our code, Energia is converting our source code in our sketch into a language that our LaunchPad's microcontroller brain can understand.

Once compiled, we can then download the code into the microcontroller brain of our LaunchPad by pressing the "arrow button" as shown in the screenshot below. This will download the compiled code into our LaunchPad brain.

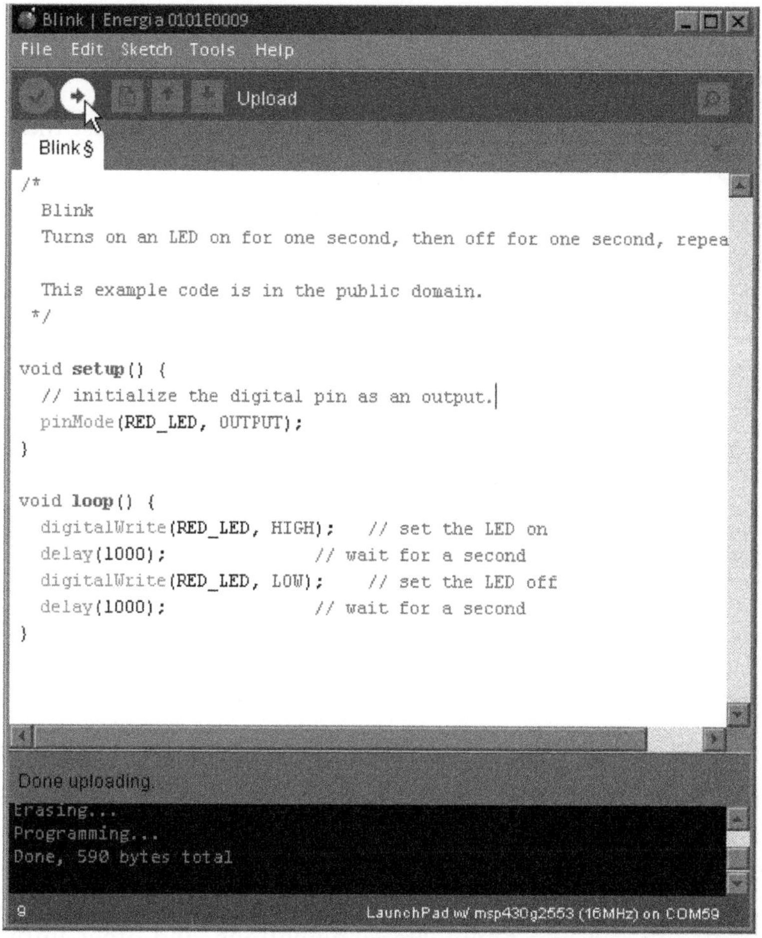

After downloading your compiled code, your LaunchPad should automatically start running the program that is now stored in the microcontroller brain. At this point, your LaunchPad should be toggling the red LED on your LaunchPad about once per second.

Also, be sure that the jumper labelled J5 (P1.0) is connected. This jumper connects the red LED to the associated pin (P1.0) on the LaunchPad.

Go ahead and pat yourself on the back! You've just become a microcontroller programmer! However, before we get too excited, let's take a deeper look at the code that we just compiled and programmed into our MSP430 microcontroller.

4.4 LOOKING AT OUR FIRST CODE EXAMPLE

Now that the LED is blinking, let's take some time to digest the code that we just downloaded into the LaunchPad brain.

```
/*
Blink
Turns on an LED on for one second, then off for one second,
repeatedly.

This example code is in the public domain.
*/

void setup() {
        // initialize the digital pin as an output.
        // Pin 14 has an LED connected on the LaunchPad
        pinMode(RED_LED, OUTPUT);
}

void loop() {
        digitalWrite(RED_LED, HIGH);    // set the LED on
        delay(1000);                    // wait for a second
        digitalWrite(RED_LED, LOW);     // set the LED off
        delay(1000);                    // wait for a second
}
```

First thing to note in this code example is the characters in green. If you are following along with your computer open, the green "stuff" in the Energia editor window that we have in our code is called *comments*, which are there to help the coder. With comments, we can add reminders, explanations, and other information to supplement the actual code and make it easier to understand. As you can see, this code example has several comments that were added to help explain what purpose each line of code has. The comments are actually ignored by Energia when it compiles the code and are only meant for the coder.

Define

Comments

Comments can be used to provide helpful tips, reminders, and explanations to supplement the actual code.

Single-line comments can be made using the following convention:

```
// A single-line comment explains what happens on that line
```

Blocks of comments can be created using the following convention:

```
/* blocks of comments can span over multiple lines, useful for
long description of a block of code or of a function*/
```

With the help of the comments, can you see what is happening? The flowchart below should also help explain our blink LED example.

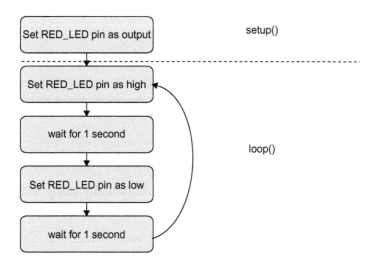

The second thing that we note is that our sketch is broken up into two specific sections called *setup()* and *loop()*, each beginning with an open bracket "{" and ending with a closed bracket "}." These big chunks of code are two examples of a function. A *function* is an identifiable, well-contained, and relatively independent block of code that performs a specific task within a larger program. The `setup()` and `loop()` functions are two specific functions that are essential to any Energia sketch and provide structure to the entire program.

Define

Function

Functions are an important concept in programming. As its name implies, a function typically has a purpose or specific action, which is either predefined or written/created by you. Once a function is created, it can be used at any time throughout the sketch to perform its specific action.

When you *create your own* function, it might look like this:

```
void functionName()
{
//perform the function here
}
```

When you want to *use* a function, whether it was created by you or precreated in Energia, you should simply type the name of the function in your code at the time that it is needed. ·

```
functionName();
```

Within the blink example sketch, we are actually using a few pre-written functions, which help us enable and control the LED.

When the sketch first starts, it calls the `setup()` function where you can initialize your LaunchPad's senses and actions. In short, we're setting up the foundation of our sketch. This is where the LaunchPad prepares the initial working environment of the sketch and readies the LaunchPad for the main loop.

Inside of `setup()`, we encounter our first function called, `pinMode()`. With the help of the comments, can you decipher what this function does? The `pinMode()` function is predefined and is inherently available inside of Energia for us to use. This is a function that is needed to configure one of the pins on the LaunchPad as an input or an output. As we learned in Chapter 2, our LaunchPad has various senses and actions at its disposal. We can use this `pinMode()` function to configure one of the LaunchPad pins to properly enable a particular sense or action. Because we are blinking an LED, we can deduce that this is a type of action or output.

Unlike `setup()` and `loop()`, `pinMode()` is a more flexible kind of function that needs additional information to help specify what it should do. This additional information is what we call *parameters*, and they are passed into the function inside of the parenthesis, separated by the commas.

Define

Parameters

Data or information that we can pass into a function as an input. These parameters are then used by the function to help it perform a specific task. A function does not always require a parameter to be provided. If multiple parameters are needed, they can be passed into a function inside of the parenthesis, separated by commas.

```
functionName(parameter1,parameter2,...)
```

In `pinMode()`, we pass in two parameters. One parameter is to let the function know which one of the microcontroller pins we want to configure. The second parameter is for the direction (input versus output) that we want to set that particular pin to. In our example, we are setting pin RED_LED as an OUTPUT.

The OUTPUT parameter makes sense as we want to perform an action; however, what does RED_LED really mean? Well, RED_LED is a nickname for pin number 2 of the MSP430G2553. Pin number 2 is connected to the red LED that we want to blink.

We can see that the red LED that we want to blink is physically connected to the microcontroller pin 2 thanks to the diagram on the next page.

LaunchPad with MSP430G2553

Revision 1.5

Energia

Hardware
Pin number
I²C
Serial UART
SPI
analogRead()
digitalRead() and digitalWrite()
digitalRead(), digitalWrite() and analogWrite()

						GROUND	20
						XIN	19	P2_6
						XOUT	18	P2_7
						TEST	17
						RESET	16
					MOSI (B0)		15	P1_7	A7
					MISO (B0)	GREEN_LED	14	P1_6	A6
			SDA				13	P2_5
			SCL				12	P2_4
							11	P2_3

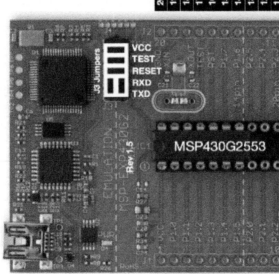

MSP430G2553

J3 Jumpers
VCC
TEST
RESET
RXD
TXD

RESET

PUSH2

TEXAS INSTRUMENTS

LaunchPad

| Flash | 16 | KB |
| Serial | | Hardware |

+3.3V			1
RED_LED	A0		P1_0	2
	A1	RXD	P1_1	3
	A2	TXD	P1_2	4
	A3		P1_3	5
PUSH2	A4		P1_4	6
	A5		P1_5	7
		SCK (B0)	P2_0	8
		CS (B0)	P2_1	9
			P2_2	10

Rei Vilo, 2012
embeddedcomputing.weebly.com
version 1.3   2012-09-09

This diagram is also available in the Appendix section.

Microcontroller pins are numbered counter-clockwise, with pin 1 being the top-left pin.

Also note that the LaunchPad silkscreen (the white print on the LaunchPad board itself) has helpful labels that explain what each pin is connected to.

Another important thing to see in the above image is that pin 14 has a few "nicknames." Pin 14 can also be referred to as:

- P1_0
- 2
- RED_LED

Energia recognizes these "nicknames" that are highlighted in the image above. For example, our code would work just the same if we were to pass "RED_LED" as a parameter instead of the number 2.

The helpful pin diagram is also printed in the Appendix: Quick References. This map will prove useful throughout our LaunchPad journey, so feel free to flip back to the Appendix as needed. The Appendix also has other useful topics and notes that will serve as a handy resource that we can leverage as we go through this book.

Here's another view of the internal connection between pin 2 and the red LED. This view is generated from a wiring and sketching program called Fritzing. We will be seeing quite a few circuit sketches similar to this throughout the book.

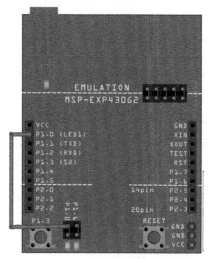

Made with **Fritzing.org**

In this particular example sketch, the setup() function is a simple one. Once the setup() function is done and the LaunchPad is all prepped up for our blink action, we enter the loop() function. The loop() function, as its name suggests, is called upon repeatedly, executing its contents over and over again. This is where the sketch spends the majority of its lifetime, performing its repetitive tasks. These tasks can be anything from gathering information from its senses, analyzing data, reacting to stimuli, or performing various actions that interacts back with the outside world. In the blink sketch, all our LaunchPad is tasked to do is turn the LED on and off every second.

In the loop() function, we encounter the next function in our example sketch called digitalWrite(). This is another function that is available to us, which allows us to set the state of a microcontroller pin as HIGH (~3 V) or LOW (~0 V). In digitalWrite(), we also pass in two parameters. Just like pinMode(), the first parameter tells the function which pin we want to talk to. The second parameter specifies which state the particular pin should be in (HIGH or LOW).

The last function that we use in this sketch is called delay(). This is another function that was prewritten for us, which causes the program to pause for a specified amount of time. This function only takes in one parameter, which is the number of milliseconds to wait for. In our example sketch, we have a delay of 1000 ms, or 1 full second.

With the help of the comments, we can see that we've taught our LaunchPad a new trick, which is to blink the red LED that is found on the LaunchPad once per second. We used the digitalWrite() function twice to do two different things for us. Thanks to its parameters, we were able to use the same digitalWrite() function to toggle pin 2 on or off. By separating those two function calls with a 1000 ms delay, we were able to create the simple blinking effect on our red LED.

4.5 MODIFYING OUR FIRST CODE EXAMPLE

Now that we understand what this blink sketch is doing, let's take some time to modify it. Here are a few exercises to test your knowledge:

Challenges:

1. As an exercise, what do you think would happen if we didn't include a delay function?

2. Try modifying this blink example to make the RED_LED toggle every 0.25 s.
3. Try modifying this blink example to make the GREEN_LED to toggle every second.
4. Try modifying this blink example to make the GREEN_LED and RED_LED blink together.
5. How about making them blink alternatingly?
6. Once you have exhausted all blinking combinations for the LaunchPad's LEDs, let's add more LEDs into the mix. Time to bust out the Educational BoosterPack. Hope you haven't forgotten about this companion. It will play an important role in our journey in upcoming chapters. For now, it can contribute four more LEDs into the mix. We'll let you figure the pin-mapping for these LEDs all by yourself. Don't forget to check http://boosterpackdepot.com/wiki/ for pin connections or refer to Appendix: *Quick Reference*.

4.6 BREAKING A PROBLEM DOWN

When you think of a computer or microcontroller system, you might think of a complex and mysterious black box made up of intricate components and involve several steps, processes, or procedures to perform complicated and mind-boggling tasks. Surprisingly enough, most microcontroller systems in the real world, like applications on the LaunchPad, are actually rather simple and monotonous!

Most microcontroller-enabled applications are designed to do a simple task repeatedly. For example, a smoke detector needs to check if smoke is present in the air over and over again. It isn't only until smoke is detected that the microcontroller needs to break from the cycle and sound an alarm. Or an alarm clock simply needs to update the time displayed on the LCD screen every second. This monotonous loop isn't broken until the current time matches the programmed alarm time and a buzzer is activated. Once you break the application down into smaller pieces, you can turn the complex problem into a sequence of simple, straightforward, and achievable tasks.

By understanding how a larger problem can be broken up into tiny LaunchPad digestible pieces, we'll be able to create the appropriate software to teach our LaunchPad how to complete the right tasks to solve many problems. As you'll soon find out, creating the operation flow and identifying these tasks is a key underlying theme for all MSP430 LaunchPad applications.

In this chapter, we introduced a lot of programming fundamentals and concepts. By understanding how the `setup()` and `loop()` functions work and learning how to leverage existing functions or even create your own, we can start solving very complex problems by breaking them down and solving each piece one at a time.

In a way, the day in the life of a microcontroller is not all that different from a normal day of a person like you or I. Much like microcontrollers, people first have to wake up from slumber and turn on. Next, people have to set themselves up for the day by brushing their teeth and putting on clothes. Once the person is set up for the day, people go to work and loop through a series of tasks, running through a monotonous and continuous flow that continues forever!

Similarly, our LaunchPad runs through a familiar flow as it executes any particular sketch. Not the most glamorous life, but someone's got to do it!

Challenges:

1. Remember the dream application that you wrote down on the first page of this book? If not, flip back a few pages to remind yourself. Once you remember, go ahead and try to describe your project idea using the `setup()` and `loop()` structure.
2. What do you think needs to happen in the `setup()` function?
3. What do you think needs to happen in the `loop()` function?
4. Draw a flowchart for your application.
5. For all actions and senses described in the flowchart, make a list of functions that you think you need. For each identified function, make a list of parameters you think each function will need to operate properly.

As you created the flowchart for the exercise above, did you have to make any decisions based on information that your LaunchPad received through one or more of the available senses? This type of decision making is important for any LaunchPad-based application.

We'll learn more about how we can teach our LaunchPad to make smart choices in the next chapter.

CHAPTER 5

Day in the Life of a Microcontroller

5.1 THE LOGICAL LAUNCHPAD

The previous chapter talks about an important concept for microcontroller and LaunchPad operation. We learned that throughout the execution of a sketch, a LaunchPad simply needs to be set up once before it can run continuously in an endless loop performing various tasks over and over again. Just imagine going through every day of your life doing the same thing over and over again without making any changes or asking any questions. That sounds rather boring!

We need to introduce a new concept to create exciting and dynamic sketches for our LaunchPad and give it a new spice of life! This chapter introduces two new essential elements of programming:

1. Conditional branch: The ability to pause and evaluate the conditions in order to ultimately make a decision.
2. Conditional loop: The ability to repeat an action or a sequence of actions based on specific conditions.

5.2 IF (HUNGRY) {EAT}

As our LaunchPad runs through its main loop, we may want to take a microsecond to pause, reflect on current conditions, and make a decision on how to proceed next based on those conditions. The simplest form of a conditional branch is the if-statement. The *if-statement* has the structure shown below:

```
if (certain condition is met) {
      perform an action
}
```

Example:

```
if (hungry) {
   eat; // in this simple example, we eat only if hungry. omnomnom
}
```

This if-statement, even though a very simple logic, allows your LaunchPad to make a decision: to choose whether or not to do a certain thing or perform a specific action.

In a flowchart, the if-statement can be represented by a diamond block, which signifies a decision point. The action flow for your LaunchPad program might look like this.

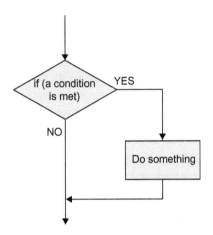

5.3 CONDITIONALS

At this point you might ask: "what constitutes a condition?". A condition is more or less a question that can be answered with either YES or NO, or in computer language terms, TRUE or FALSE. Within Energia, we have several tools or operations to help us find the answer to these TRUE or FALSE questions.

Our LaunchPad brain is typically working with either numbers or booleans (TRUE or FALSE). Using operators, we can test to see if numbers are within a particular range, or if something is true or false.

Here's a list of some of the acceptable arithmetic comparison operators available to us:

```
==   equal
!=   not equal
<    less than
<=   less than or equal to
>    greater than
>=   greater than or equal to
```

Using the above operators, we can test the condition at any of these decision points and take action depending on whether or not that particular condition is met. To help drive this home, let's go to the LaunchPad headquarters at Texas Instruments in Dallas—more specifically in the summer time. As you can imagine, the Dallas summer is very hot and the nifty if-statement and the available arithmetic operators help to keep Texans cool!

```
void(){
        if (temp >= 70){
                airCondition(on);     // based on Texas summers
                                      // the A/C will always on!!
        }
        if (temp < 70){
                airCondition(off);
        }
}
```

In addition to testing individual conditions, we can also test for multiple conditions within a single if-statement through the use of logical operations:

OR (||)   reads TRUE if at least one of the conditions are met.
AND (&&)   reads TRUE only if all conditions are met.

Thanks to the help of logical operands, we can test for multiple conditions inside of an if-statement as we show below in our modified air conditioning example:

```
void(){
    if(temp >= 70){
      airCondition(on);     // based on Texas summers, our
                            // A/C is always on!!
      heater(off);
    }
    // if temp is within 60 and 70, the weather is perfect! Both
    // heater and A/C is off
    if(temp >=60 && temp < 70 ){
      airCondition(off);
      heater(off);
    }
    // if temp is lower than 60, it's too cold! Turn that heater on!
    if(temp < 60){
      airCondition(off);
      heater(on);
      }
    }
```

Challenge:

Can you think of three conditional activities that you perform daily? For each activity, what triggers you to do it? Can you shape this into the form of an if-statement, if (some condition is met) {do this action}?

5.4 IF (HUNGRY) {EAT} ELSE {SLEEP}:

A more flexible version of the if-statement offers you an alternative choice, allowing you to either do something if the condition is met, or do something *else* otherwise. This is what we call the if-else-statement.

In the syntax form, it might look like the structure shown below.

```
if (certain condition is met) {
        perform action A
}
else
{
        perform action B
}
```

Example:

```
if (hungry) {
        eat;      // omnomnom
}
else{
        don't eat, go to spleep;  // i'm full, zzzzzzzzz...
}
```

In a flowchart, the if-then block might be used in a way similar to the block diagram below.

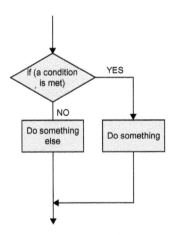

> **Note**
>
> If you now compare the two flow diagrams of the if-statement and the if-else-statements, you will realize that `if()` is simply a special case of the `if() else` where your LaunchPad simply does not do anything in the event the else branch takes place.

Let's revisit the temperature example, but this time using an if-else-statement.

In our first attempt at being an HVAC (heater, ventilation, air conditioning) programmer, we had to perform three independent if-statements to check for three ranges of temperature in Fahrenheit (less than 60°F, between 60°F and 70°F, and above 70°F). However, with the if-else-statement, we can actually make some improvements to our original code example.

```
if(temp >= 70){
    airCondition(on);// based on Texas summers, the A/C is always on!!
    heater(off);
}
else { // if we end up in the else section, it means that temp is < 70
       // need to check the other temp ranges
       // what do we need to do to check the other temp ranges?
    }
```

In this first if-else-statement, we check to see if temperature is $>= 70°$F. If not, we jump to the else branch to check the other temperature ranges. If we branch to the else portion of this if-else-statement, we know that temperature has to be $<70°$F, otherwise we would not have branched there.

If we branched to the *else* portion of the if-else-statement, we next need to check the other temperature ranges. In order to do this, we're actually going to need a second if-statement that will check for the next temperature range (`temp >= 60 && temp < 70`). However, because we already know that `temp <70`, we can remove this portion of the condition and this simply becomes `if (temp >60)` as shown below.

```
if(temp >= 70){
  airCondition(on); // based on Texas summers, our A/C is always on!!
  heater(off);
}
else { // if we end up in the else section, it means that temp
      // is < 70
       if(temp >= 60 ) { // We already know it's <70, so we only
                         // check if temp >60
  airCondition(off); //both heater and A/C is off
  heater(off);
       }
}
```

Wow, an if-statement inside another if-statement? This is called a *nested-if-statement*. But fear not, you can always trace your steps and know where you are if you pay attention to the friendly and helpful curly braces {}. As you recall, they group a block of code together in a nice unit. If you want to find out which if- or else-statement your code is inside of, simply move your cursor to one of the curly braces and click on it. Energia will help you find the curly brace's corresponding mate and highlight it for your convenience. Neat trick, huh?

Back to our temperature situation. So far so good, but what about the last condition for when temperature is <60°F? Well, if you trace through the if-statements so far, in the event that temp <60, we will need an else-statement to capture the condition for when temperature is <60. We show the complete example below with our nested-if-statements!

```
if (temp >= 70) {
    airCondition(on); // based on Texas summers, A/C is always on!!
    heater(off);
}

else { // if we end up in the else section, temp is < 70
    if (temp >= 60) { // We already know it's < 70,
                      // so we only check if temp >= 60
        airCondition(off); // both heater and A/C are off
        heater(off);
    }
    else { // brrrrrr, temp < 60
        airCondition(off);
        heater(on);
    }
}
```

With the if-else-statement, we enable our LaunchPad to test against multiple conditions and take different courses of action depending on whether a particular condition is met or not. Additionally, the ability to nest if-statements within each other enables unique possibilities and further intelligence for your LaunchPad! Aren't you so proud of your LaunchPad? It's getting smart!

5.5 MULTIPLE CHOICES WITH THE SWITCH

As powerful as our if-else-statement is, sometimes two choices are not enough. After all, the world as we know it is not simply black and white. If the LaunchPad needs to make a single choice out of multiple options, we can use yet another conditional structure that we call the *switch-statement*.

```
switch (variableX) {
      case valueOne: // execute this case if variableX = valueOne
            do Action_1;
            break;
      case valueTwo: // execute this case if variableX = valueTwo
            do Action_2;
            break;
      case valueX: // execute this case if variableX = valueX
            do Action_x;
            break;
      default:    // execute this default case if none of the
                  // above cases are satisfied
            do Action_??;
            break;
}
```

This switch-statement can be summarized by the flow diagram below.

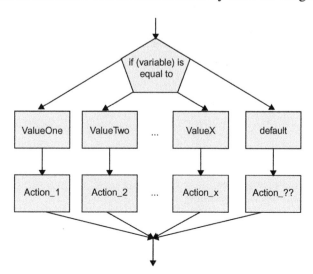

Here's a simple example of how we can apply the switch-statement into something that we can definitely relate to. Based on time, we can choose multiple actions. In this case, we are sleeping all day, except waking up at 7AM, noon, and 7PM to eat breakfast, lunch, and dinner.

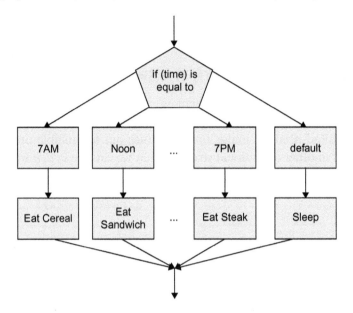

We will be using this switch-statement in future projects inside of this book! The switch-statement is a very organized way of adding intelligence to your LaunchPad, especially when there are multiple conditions that the LaunchPad is waiting for to be satisfied.

One thing that you may note is that our switch-statement is actually a glorified if-else-else-else...statement that checks against multiple conditions. As you can see in the structure below, the switch-statement is identical to the following if-else-else-else...statement.

Example:

```
if(time == 7AM){
        Eat Cereal; // omnomnom
}
else if (time == noon){
        Eat Sandwich; // omnomnom
}
else if (time == 7PM){
        Eat Steak; // omnomnom
}
else{
        Sleep; // zzzzzzzzz.....
}
```

Challenges:

1. Thinking back to your dream application that we wrote down in the beginning of the book, do you think you can leverage any of these conditional branches?
2. Can you think of other possible scenarios that may benefit from the use of a `switch()` statement?
3. Can you think of a possible scenario that may benefit from a nested-if-statement?

5.6 REPETITION + CONDITION = LOOP

Another fundamental aspect of controlling the flow and making decisions for your sketch is the ability to repeat a certain action or group of actions. Sometimes you need to repeat an action for an exact number of times, while in other occasions you might need to repeat the action(s) until a certain condition is met. This is where conditions can be used to decide on the number or duration of repetitions. Combining the two elements of repetition and condition gives us the programming concept of a conditional loop.

A *conditional loop* can be described as a group of actions being repeated as long as a certain condition is met. The first loop keyword we want to introduce to you is **while**. Using pseudocode, here's what a **while** loop looks like.

```
while (a condition is true){
      perform this action
}
```

or back to our hungry example:

```
while (still hungry) {
      take a bite
      // in this example, we're going to keep taking a bite
      // as long as we're still hungry
}
```

The loops mentioned above will continuously run as long as the particular conditions inside of the parenthesis are true. A while loop is particularly useful when we don't necessarily know how many times we need to execute a loop. This type of loop is useful in many cases, such as in our HVAC example in the previous section. Let's pretend it is another hot summer day in Texas...

```
while (temperature > 70) {
    airCondition(lower); // crank up the AC another notch
}
```

In the example above, we use the while loop to continuously lower the temperature until temperature is $< = 70°F$. Once temperature is $< = 70°F$, the conditions inside of the parenthesis are no longer true and the loop stops. In this particular example, we don't want to hard-code the amount of loops because we don't necessarily know how many iterations of `airCondition(lower)` we will need to reach our desired temperature of $70°F$.

In other situations, you might know in advance exactly how many times you would like the loop to repeat itself. In this case, the particular loop that works best for you is the for-loop. The structure of a for-loop looks a tiny bit different compared to the while loop, so let's start off with an example and we can dissect the code to see what's going on. Let's say you want to have 10 slices of cake (don't judge us, we like to eat...a lot).

```
for (int counter = 0; counter < 10 ; counter++) {
// the for-loop accepts 3 clauses, which are separated by
//semicolons
    eat a slice of cake;
}
```

As you can see in the example above, the for-loop accepts three clauses:

- The first clause tells the for-loop which variable to pay attention to. As shown in the example, we can actually initialize and declare a new variable specifically inside of the for-loop. In this case, we made it an integer variable called "counter."
- The second clause is the condition that we will test for before the execution of a loop. If that particular condition is true, we will execute the loop one more time.
- The third clause tells the for-loop what to do at the end of each loop iteration. `counter++` is actually the same as saying `counter = counter + 1`. It simply increments the counter by 1.

So in our above example, our counter is initialized at 0 slices of cake. As it's initialized at 0, counter is definitely less than 10. Since we satisfy the condition, we execute the loop at least once and eat a slice of cake! Yum! After each iteration of the loop, we told the for-loop to increment the counter, so counter now equals 1.

Since 1 is still less than 10, we go ahead and execute the loop again and eat another slice of cake. And just like the previous iteration, we increment our counter to 2.

We continue to do this until we increment counter 10 times. Once counter = 10, we no longer satisfy the condition of our for-loop because 10 is not less than 10. At that point, we exit the loop and do not run through any additional iterations.

The key ingredient of a for-loop is the use of a running counter that starts off at the initial value, ends when it reaches the end value, and increments each iteration of the loop.

5.7 PROJECT 3: LED BLINK COUNTER

Let's apply the for-loop to our blinking LED example to get a better idea of how to use this. Let's say we need to blink the LED exactly 10 times. As each blink consists of turning the LED on and off, we have to do both actions 10 times.

Ingredients:

1. LaunchPad

In this project, we will be using the onboard green LED of the LaunchPad, which is tied to pin 14. Because we are trying to blink the green LED, be sure that the jumper labelled J5 (P1.6) on the LaunchPad is connected.

```
void setup() {
        // initialize the digital pin as an output.
        pinMode(14, OUTPUT);
}

void loop(){
        for (int i = 0; i < 10; i++){ // blink the green LED 10x with
                                       // the help of the for-loop!
             digitalWrite(14, HIGH);
             delay(300);
             digitalWrite(14, LOW);
             delay(300);
            }
            // when we're done with the for-loop, exit the
            // loop and continue the sketch
            while(1);
            // once we blinked the LED the desired amount of times,
            // we get stuck in this empty while loop
            }
```

Compile, download, and execute this code in Energia. We should see our green LED blink exactly 10 times! Thanks to our for-loop, we ensure that the LED blinks for the specified amount of times.

We also introduce a new concept—the infinite loop. At the end of our sketch, we have an empty while loop that is always true. This means that this while loop does nothing forever! The reason we add this to the end of our code is so that our main loop() only executes once. If we did not have the while(1) function there, the LED would blink 10 times over and over again.

In addition to the infinite loop, there is another thing we can do to ensure that the LED only blinks the appropriate amount of times. If we want to make sure the LED only blinks 10 times, we can move this block of code into the setup() function, where it only gets to run once. At this point, nothing will be left in the main loop() function. Your final version of the code should look something like this.

```
void setup() {
        pinMode( 14, OUTPUT);

        for (int i = 0; i < 10; i++){
                digitalWrite(14, HIGH);
                delay(1000);
                digitalWrite(14, LOW);
                delay(1000);
        }
}

void loop () {
        // Do nothing, setup() already blinks the LED 10 times.
}
```

One More Thing...

The for-loop isn't the only way to repeat a block of code for an exact number of times. You can actually do this with a while loop as well. Can you figure out how to do this?

Hint: You will need to create your own counter and use that counter in the condition for the while loop.

```
int i = 0;      // this while loop operates similarly to the
                // for-loop example previously.
while(i<10){    // the while loop will only loop 10 times and blink
                // the LED only 10 times.
        blinkLED();
        i++;
}
```

As a matter of fact, under the hood, this is how a for-loop is constructed using a while loop.

In a for-loop, the counter does not always necessarily have to count up or increment by one. The following loops are also valid.

```
for (int i = 10; i > 0; i--) {
// counter i starting off at 10, decrementing by 1 each time
// until 0.
// do something 10 times

}

for (int k = 0; k < 20; k=k+2) {
// counter k starting off at 0, incrementing by 2 each time
// until 20.
// do something
// still 10 slices of cake though, we're not .... that greedy
}
```

5.8 YOUR LAUNCHPAD JUST GOT EDUCATED!

This chapter introduced a lot of important life-changing concepts for our LaunchPad! And just like for you and us, making a string of good decisions can lead to a successful and happy LaunchPad.

Now that you understand how we can insert logic into our LaunchPad using conditional if-statements, switch-statements, and loops, our LaunchPad can now make intelligent decisions based on the information gathered from its many senses.

Using these simple and fundamental logic blocks, our LaunchPad can control many things. The simple question of "if" opens up a brave new world for our little LaunchPad, a world that paints many different paths on the journey to conquer the LaunchPad quest. And by every decision that it makes, the LaunchPad takes yet another step closer to the final destination, reflecting on the splits it encountered, the roads it took, and the roads not taken.

CHAPTER 6

Think Digitally

The microcontroller world is a digital one. Digital usually refers to discrete, noncontinuous information. We've touched on a few examples already—true or false, yes or no, high or low, and one or zero. These datasets are discrete, in that it's either one or the other. This type of data is what we call digital.

The brain of our LaunchPad can only process digital data and handle this digital information within a series of bits. For our MSP430 LaunchPad, the brain is a 16-bit processor. This means that most of its operations, datasets, and instructions are done 16-bits at a time. Each of these bits can either represent a "1" or a "0." It's pretty amazing what we can do with just these two numbers!

6.1 BINARY: 0 + 1 = 65,536

In a 16-bit strand, we can have many combinations of 1s and 0s, which can represent various individual values. To be exact, we can use a 16-bit strand of either 1s or 0s to represent 65,536 different values! This type of data format is called *binary*. While each bit can only carry either a 1 or a 0, we can use a unique strand of 16-bits to represent any number we want! Below are a few examples:

Binary   Decimal
$0000_2 = 0000_{10}$
$0001_2 = 0001_{10}$
$0010_2 = 0002_{10}$
$0011_2 = 0003_{10}$
$0100_2 = 0004_{10}$
$1000_2 = 0008_{10}$
$1001_2 = 0009_{10}$
$1010_2 = 0010_{10}$
$1111111111111111_2 = 65,536_{10}$

> **Note**
>
> The subscript denotes the base of the numeric system. In the case of binary, the base is 2. In decimal, the base is 10.
>   So how do we go from binary to the more familiar decimal format? Well, the first thing we should do is examine how our traditional decimal (base 10) format works...

In decimal format, we have multiple "places." We have a "ones" place, a "tens" place, a "hundreds" place, a "thousands" place, and so on and so forth. This is very similar to binary, except that we have a "ones" place, a "twos" place, a "fours" place, an "eights" place, and so on and so forth.

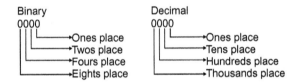

In decimal, each "place" can hold a digit between 0 and 9. In binary, however, each "place" can only hold either a 1 or 0. Before we give you the "formula" to convert from binary to decimal, let's think about this a little more.

To count up in decimal, we start out in the ones place, counting from 0 to 9. But in order to get to 10, our ones place resets to 0, and we have to carry a digit over to the tens place. Once we carry over a digit into the tens place, we can continue to increment back in the ones place—11, 12, 13, 14....

Now, let's try to apply this logic in binary. To count up in binary, we start in the one's place, but we can only count from 0 to 1. In order to get to 2, we need to carry a digit over into the twos place "and reset the digit in the ones place. Once we carry a digit over to the twos' place", we can then start incrementing within the ones place again. Also, just like in the decimal format, there are times when the digit we have to carry over propagates over multiple places. For example, when we increment from 999_{10}, the extra digit carries all the way over to the thousands place. This is the same case if we are to increment from

111_2 in binary (this is equal to 7_{10} in decimal). To increment from 111_2, we have to carry the extra digit all the way into the eights place.

Phew! That wasn't so bad! Now that we have gone through the concept, let's take a look at a formula we can use to convert from binary to decimal.

To show you how to convert from binary to decimal, let's try an example.

Challenge:

Convert this binary to decimal: 11001001_2.

To make this conversion, we have to do a "weighted sum" for each place. As you can see, the weight of each place is determined by its position. The rightmost digit has the smallest weight. The leftmost digit has the largest weight. We can also see that the weight of each place is based on an exponent of the base number. The base number for binary is 2.

1	1	0	0	1	0	0	1
$2^7 = 128$	$2^6 = 64$	$2^5 = 32$	$2^4 = 16$	$2^3 = 8$	$2^2 = 4$	$2^1 = 2$	$2^0 = 1$

By doing our weighted sum, we get the following to convert from binary to decimal.

$$1 \times 128 + 1 \times 64 + 0 \times 32 + 0 \times 16 + 1 \times 8 + 0 \times 4 + 0 \times 2 + 1 \times 1 = 201_{10}$$

This is actually very similar to how decimal works! Let's take a look at a base 10 number system.

Let's look at the decimal number 1583 as an example.

1	5	8	3
$10^3 = 1000$	$10^2 = 100$	$10^1 = 10$	$10^0 = 1$

$$1 \times 1000 + 5 \times 100 + 8 \times 10 + 3 \times 1 = 1583$$

Just like our binary example, each place is weighted. The weight is again determined by an exponent of the base number. In a decimal number system, the base is 10.

It checks out! Note that the binary and decimal numbers operate very similarly. The only difference is that in binary, each place can only accept two digits (0 or 1), while in decimal, each place can only accept ten digits (0 to 9). The other difference between different number formats is that the base number (2 in binary, 10 in decimal) dictates the "weight" of each digits place, which is calculated based on an exponent.

6.2 HEXADECIMAL—BINARY'S BIG BROTHER

Hexadecimal is another way of representing numbers. Hexadecimal is an interesting numbering scheme, which counts from 0 through F. Let's take a closer look at what we mean.

This table summarizes hexadecimal, binary, and decimal format.

Decimal (Base 10)	Binary (Base 2)	Hexadecimal (Base 16)
0	00000011	0x0
1	00000001	0x1
2	00000010	0x2
3	00000011	0x3
4	00000100	0x4
5	00000101	0x5
6	00000110	0x6
7	00000111	0x7
8	00001000	0x8
9	00001001	0x9
10	00001010	0xA
11	00001011	0xB
12	00001100	0xC
13	00001101	0xD
14	00001110	0xE
15	00001111	0xF

Note

While binary numbers usually stand out having only 0s and 1s, hexadecimal numbers sometimes might look like decimal numbers when A−F are not used. To identify a number as a hexadecimal value instead of a decimal value, a 0x-prefix can be added as shown in the table above.

Hexadecimal format is used in lots of microcontroller code because of the way many microcontrollers handle and manipulate data. Most microcontrollers work with data 8-bits at a time. This is what we call a *byte*. We typically represent a byte of data in either a string of eight 1s or 0s, or as a 2-digit hexadecimal value. To convert from binary to hex, we simply split an 8-bit binary number into two sets of 4-bits. Once we've split it into two 4-bit values, we can determine the appropriate hex value.

Example

$1001\ 1110_2 \rightarrow 9E_{16}$

Challenges:

Convert the following binary numbers into decimal! To check if you're right, flip to that page in this book!

- 01010111_2
- 00101111_2

Convert the following hexadecimal numbers into decimal! To check if you're right, flip to that page in this book!

- 0x74
- 0x63

6.3 WHAT'S WITH THE MATH CLASS?

OK—sorry, you're right! Let's get back to blinking some LEDs! However, the binary system is extremely important for understanding how microcontrollers handle digital data, so I'm glad you took the time to read through it! Now that we know that our LaunchPad brain can really only handle 1s and 0s, it's pretty amazing what we can do with just those two numbers. For example, with either a 0 or a 1 connected to the right bit, we can turn an LED off or on.

●●●

Tip

Use Windows calculator in programmer's mode for binary and hex conversions.

6.4 DISSECTING THE LAUNCHPAD'S BRAIN

Our LaunchPad's brain is based on a microcontroller from Texas Instruments called MSP430. The specific MSP430 device on the LaunchPad is the MSP430G2553. This device is an embedded processor, which can handle millions of 16-bit instructions per second. Amazing, right?

By flipping 1s and 0s extremely fast, our LaunchPad brain is able to make decisions, control things, blink LEDs, spin motors, and more.

Inside of our MSP430 device on our LaunchPad is a central processing unit, or CPU. It is a simple unit that takes in 1s and 0s, processes them, then spits out a new batch of 1s and 0s. Also inside of our LaunchPad brain are what we call peripherals. These peripherals are integrated into the LaunchPad brain to provide different types of functionality. One of the integrated peripherals is a General Purpose Input/Output (GPIO) module.

This integrated GPIO module is what provides the various pins that poke outside of the MSP430G2553 device. As we've learned, these pins can be configured as inputs or outputs and can carry a value of HIGH or LOW.

In addition to the GPIO module, the MSP430 brain on our LaunchPad has other peripherals as well, including analog-to-digital modules, serial communication modules, and more! We'll get the chance to play with each of these peripherals in later chapters!

The reason we are introducing these various peripherals that are inside of our LaunchPad is because they are controlled by these 1s and 0s that we keep talking about.

Inside of our LaunchPad brain, we have a collection of *registers*. A register is a holder of bits inside of our LaunchPad brain. They are capable of holding 1 byte (or 8-bits) of data at a time.

Each of the peripherals inside of our LaunchPad, including our GPIO module, are controlled by one or more register. And based on the 1s and 0s that populate any given register, our peripherals may act or behave in a different manner.

As an example, our GPIO module on our LaunchPad is controlled and defined by a few registers. However, the main registers that control our GPIO module is an 8-bit register called P1DIR and and another 8-bit register called P1OUT.

We can populate these registers with different 1s or 0s to make our GPIO module do different things. We'll play around with this idea in yet another blink LED project.

Looking at the MSP430G2553 documentation, we can learn all about the different registers available inside of the LaunchPad brain. Here is what the documentation says about the registers P1DIR and P1OUT that we will be using.

Direction Registers PxDIR

Each bit in each PxDIR register selects the direction of the corresponding I/O pin, regardless of the selected function for the pin. PxDIR bits for I/O pins that are selected for other functions must be set as required by the other function.

Bit = 0: The port pin is switched to input direction.
Bit = 1: The port pin is switched to output direction.

The "x" in PxDIR is a placeholder for the port number that we want to configure. Because PxDIR is an 8-bit register, we have 1 bit for each pin that is available on the port. To set a specific pin as an input, we set the corresponding bit in the register LOW (0). To set a specific pin as an output, we set the corresponding bit in the register HIGH (1). So, for example, the register P1DIR is the register that controls the pins available on port 1. We can set each bit as a 1 or 0 to make that particular pin an output or an input as shown below.

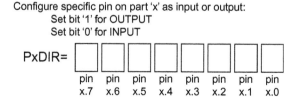

Configure specific pin on part 'x' as input or output:
Set bit '1' for OUTPUT
Set bit '0' for INPUT

PxDIR=

pin x.7 pin x.6 pin x.5 pin x.4 pin x.3 pin x.2 pin x.1 pin x.0

We need to tell this register which pins are inputs and which are outputs so that the LaunchPad brain knows what to do with each pin. For example, if we want to set pin P1.6 as an *output* (this is the pin our Green LED on LaunchPad is connected to, by the way) we simply set P1DIR = 01000000_2.

The other register that we will introduce to you is PxOUT. This register is also 8-bit wide and the 1s and 0s that we place into it will set the state of the port pin. Here is what the MSP430G2553 documentation says about the PxOUT register.

Output Registers PxOUT

Each bit in each PxOUT register is the value to be output on the corresponding I/O pin when the pin is configured as I/O function, output direction, and the pullup/down resistor is disabled.

Bit = 0: The output is low.
Bit = 1: The output is high.

Just like PxDIR, the x corresponds to the specific port that we want to control. Also like PxDIR, PxOUT is an 8-bit register where each bit corresponds to a specific pin. We use 1s and 0s to make the signal that is being output on a specific pin HIGH or LOW.

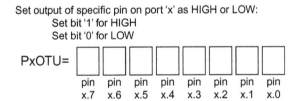

6.5 PROJECT 4: CONTROLLING THE LEDS THE HARD WAY!

In this project, we will learn about how our LaunchPad brain works. As mentioned, the Launchpad brain is integrated with different peripherals that offer different capabilities. Each of these peripherals are controlled by registers that are inside of the LaunchPad brain. By putting in different combinations of 1s and 0s into these peripheral registers, we can change what each peripheral does and how it behaves. For this project, we are going to control the Green LED on our LaunchPad by directly modifying the registers for the GPIO module.

Ingredients:

1. LaunchPad

To set up the hardware for this project, simply plug your LaunchPad into the USB port of your computer. Once connected, also make sure the jumper is connected for the green LED. Once your LaunchPad is ready to go, let's turn our attention to Energia and write some code.

```
void setup() {
    // pinMode(GREEN_LED, OUTPUT);
    P1DIR = 0b01000000; // 0x40 in hex

    // digitalWrite(GREEN_LED, OUTPUT);
    P1OUT = 0b01000000; // 0x40
}
void loop() {
}
```

This code example simply configures the appropriate port pin as an output and turns the LED on by setting the port pin into a HIGH state. Also, we commented out the `pinMode()` and `digitalWrite()` functions that we used in a previous project and replaced them with new code that directly modifies the registers of our GPIO module to accomplish the exact same thing.

To use binary in Energia, we simply add "0b" before the binary string.

Alternatively for hex numbers inside of Energia, we add "0x" to the beginning of the hexadecimal value.

What we are seeing here is the "register level" of microcontroller programming. At the register level, microcontroller developers can have lots of flexibility and control over the LaunchPad. However, this extra flexibility comes with extra complexity as we are responsible for flipping the specific 1s and 0s in meaningful ways inside of the appropriate registers. Luckily for us, the functions that Energia uses (including pinMode, digitalWrite, and more) hide the lower level registers from programmers. Instead of typing 1s and 0s, we are able to use friendlier functions that are easier to use and understand.

Compile, download and run the code above. Try the code first with binary. When that works, try out hexadecimal in the code instead. Because the hexadecimal and binary numbers are representing the same value in different formats, we should get the exact same functionality regardless of which number format we decide to use.

The main thing to take away here is that the inside of our LaunchPad brain is a constantly changing landscape of 1s and 0s. Fortunately for us, the friendly function calls that Energia uses hides away the peripheral registers and the 1s and 0s inside of each register.

Challenges:

1. Try to configure the RED_LED as an output and turn it on using binary with our P1DIR and P1OUT registers.
2. Try to configure the RED_LED as an output and turn it on using hexadecimal with our P1DIR and P1OUT registers
3. Try to use make the LED blink by leveraging the code we used in a previous project while using P1DIR and P1OUT registers.

6.6 DIVING DEEP TO THE REGISTER LEVEL

So why did we just learn how to blink LEDs the hard way? While we won't be twiddling bits and bytes at the register level for the rest of the book, it's a good exercise to understand what is happening behind the easy-to-use function calls that Energia provides us. The more user-friendly functions are great because they allow the developer to focus on solving the problem at hand and not have to worry about the underlying bits and registers.

Every once in a while, developers may require the added flexibility of the lower level registers to optimize their particular application. What we gain in ease-of-use, we lose in the ability of full optimization by abstracting away from the low-level registers by way of higher-level function calls.

CHAPTER 7

The Ins and Outs of Digital I/O

At this point, you have learned that our LaunchPad has a few senses and actions available to it. Our sketches enable the LaunchPad to perform various actions based on the information gathered through its senses. As we learned in Chapter 6, our LaunchPad can interface with external objects through its many pins. These pins can be connected to buttons, LEDs, buzzers, and other external components.

In this chapter, we will learn about the digital input and output signals that our LaunchPad can use to interact with the outside world.

7.1 PROJECT 5: READING A BUTTON INPUT

Let's go back to Energia to play with this concept first hand. In this project, we will play with both inputs and outputs by changing the state of an LED (output) by pressing a button (input).

Ingredients:

1. LaunchPad

To start things off, let's load up an example sketch called "button." We can find this example sketch by following the path below in Energia:

File > Examples > 2. Digital > Button

```
const int buttonPin = PUSH2; //the number of the pushbutton pin (P1.3)
const int ledPin = GREEN_LED; //the number of the LED pin (P1.6)

int buttonState = 0; // variable for reading the pushbutton status

void setup() {
        // initialize the LED pin as an output:
        pinMode(ledPin, OUTPUT);
        // initialize the pushbutton pin as an input:
        pinMode(buttonPin, INPUT_PULLUP);
}
```

```
void loop() {
        // read the state of the pushbutton value:
        buttonState = digitalRead(buttonPin);

        // check if the pushbutton is pressed. If pressed,
        // buttonState = 0 (LOW)
        if (buttonState == HIGH) {
                digitalWrite(ledPin, HIGH);
        }
        else {
                digitalWrite(ledPin, LOW);
        }
}
```

Compile and download this code into your LaunchPad. This particular sketch does a good job of introducing us to the concept of I/O. In this case, a button acts as our input while the green LED serves as our output. Press Switch 2 on your LaunchPad. Does the green LED react the way you think it should, based on the code above? By default, the green LED should be on. When the button is pressed, the green LED should turn off. Be sure that the jumper J5 is populated so that the green LED is connected to the MSP430 port pin.

7.2 CONFIGURING I/O

Let's walk through the sketch line by line. First, we have to do a few things inside of our setup() function.

```
const int buttonPin = PUSH2; // the pin number of the pushbutton pin (P1.3)
const int ledPin = GREEN_LED;// the pin number of the LED pin (P1.6)

int buttonState = 0;          // variable for reading the pushbutton
                              // status

void setup() {
        // initialize the LED pin as an output:
        pinMode(ledPin, OUTPUT);
        // initialize the pushbutton pin as an input:
        pinMode(buttonPin, INPUT_PULLUP);
}

//....
```

As we learned in Chapter 4, we set the foundation for the rest of our sketch inside of the setup() function. In this case, we have to set up our input and outputs, or senses and actions. First thing we do is set up our LED as an output pin. We do this by using our pinMode() function, which we were first introduced in Chapter 4. The `pinMode()` function can also be used to set up our input, which is the pushbutton. Based on the parameters that we pass into the function, we can configure our microcontroller properly.

For the most part, this is very similar to our original blink sketch from our first project. There are a few differences, however. We have a few lines of new code before the **setup**() function, and we also introduce a new parameter to our `pinMode()` function, INPUT_PULLUP. First, let's learn about those lines of code that we have outside of the **setup**() function.

From the pin-mapping guide in the Appendix section, we can see that the green LED (aka GREEN_LED) is tied to pin 14 (P1.6) while the pushbutton (aka PUSH2) is tied to pin 5 (P1.3). Note, however, that while it would work, we did not pass GREEN_LED or PUSH2 as a parameter into the pinMode functions in the code example. Instead, we passed in something completely different called "buttonPin" and "ledPin." These are what we call "variables."

We can think of variables as data holders. These variables can be predefined or created by us, the coder. An example of a predefined variable is GREEN_LED. GREEN_LED is a variable that is inherently available inside of Energia and represents the number 14, which signifies the microcontroller pin number that the green LED is connected to. Similarly, PUSH2 is another inherent variable predefined inside of Energia and represents the number 5, which is the pin number that connects to the button. As you can see in the pin-mapping guide, the brain of our LaunchPad has 20 pins. And in this example, we're using helpful variables to symbolize the actual pin numbers that we need for this simple project.

In this example, however, we created a few user-defined variables at the top of our sketch before the **setup**() function. The two variables that we created are called *buttonPin*, which we set equal to PUSH2 (5) and *ledPin*, which we set equal to GREEN_LED (14). As we glance through the rest of the sketch, we can see that we use the new variables as opposed to directly using PUSH2 and GREEN_LED every time. Why, you ask?

The reason that this particular sketch uses the new variables throughout the code is so that this code example can be reused and modified easily. In the case of our LaunchPad, our input, or button switch is always connected to pin 5. Also on our LaunchPad, the green LED is always connected to pin 14. However, what if we wanted to

create a different project that has a button and an LED that needed to be tied to different port pins? Thanks to the two variables that we created, all we have to do is modify the value of buttonPin and ledPin once at the top of our code, and the new button and LED pin assignments would propagate through the rest of the code! Had we not created these new variables that held the pin assignments for our button and LED, we'd have to go through our entire sketch and update each pin assignment manually. Sometimes it pays off to be a little lazy!

Another thing to note is that we set these variables as a *constant* when we first initialize them. We can do this by typing "const" before the variable definition. This means that once defined, these variables cannot be changed and remain constant while this code is executing and running in the LaunchPad. This makes sense because the LED and button are not going to change pin assignments during the course of the sketch.

7.3 HOW A BUTTON WORKS

So now we know why we pass ledPin and buttonPin as parameters into our pinMode functions. Do you notice anything else different? We can see that the pinMode function that we are using to initialize our button is using a unique parameter called INPUT_PULLUP. The first part "INPUT" makes sense because this button serves as a "sense" or input to our LaunchPad brain, but what does the "PULLUP" part mean? Well, now things get interesting! To find the answer to this question, we have to understand how a button or switch works.

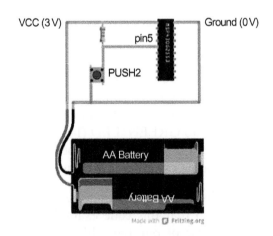

Above is a schematic, or hardware design, of how our button is connected to our microcontroller. As you can see, the button is connected to pin 5, which is exactly the value of the parameter that we passed into `pinMode()`. We can think of a button the same way as we think of a switch. A switch can be left open or closed.

When a switch is closed, it completes a circuit and allows electricity, more specifically electrons, to flow. We won't get into too much detail here, but the main thing to remember is that a switch helps us control the flow of electricity, which we use to send input and output signals to and from the microcontroller.

Let's examine this button circuit in two different scenarios—when the switch is pressed versus not pressed.

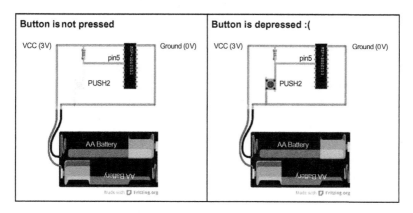

What do you think happens to pin 5 when the button is pressed? What about when the button is not pressed?

When the button *is not* pressed, we can see that pin 5 is connected to 3 V (VCC) by default thanks to our pull-up resistor.

The pull-up resistor shown in the images above enables a default electrical connection between pin 5 and 3 V. It's also important to note that pull-down resistors can also be used, which defaults a pin connection to ground, or 0 V. In this case, however, we are using a pull-up resistor.

Define

Pull-Up Resistor
A pull-up resistor enables a default hardware connection for a pin to set it to VCC, 3 V.

Note that pull-down resistors can also be used, which defaults a hardware connection to ground, 0 V.

When the button *is* pressed, we close the circuit and connect pin 5 to ground, resulting in 0 V being connected to pin 5.

We just introduced an important concept in how electrical signals correlate to input and output signals for our microcontroller. 3 V in this case correlates to a "HIGH" signal, or a "1." 0 V on the other hand correlates to a "LOW" signal or a "0." It is through electrical signals where our LaunchPad can interface with the outside world with digital inputs and outputs.

So how does this all relate back to our `pinMode()` parameter `INPUT_PULLUP`? Well, the MSP430 microcontroller that is found on our LaunchPad has special resistors integrated inside of the microcontroller itself! By passing the `INPUT_PULLUP` parameter into our `pinMode()` function, we are actually enabling an internal pull-up resistor that is available inside of the microcontroller brain. This pull-up resistor, when enabled, makes the default electrical connection between VCC (3 V) and pin 5. Thanks to the pull-up resistor, we are guaranteed to see a HIGH signal on pin 5 when the button is not pressed. Let's think about what might happen if we simply passed in `pinMode(buttonPin, INPUT)` instead of `pinMode(buttonPin, INPUT_PULLUP)`. It would result in the following circuit.

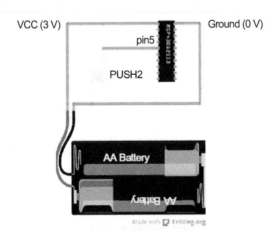

By not enabling the pull-up resistor in our `pinMode()` function, pin 5 is left in an unknown state when the button is not pressed. In this scenario, the pin is called to be "floating." At this point, we cannot be sure what signal is being received by pin 5. By enabling the pull-up resistor, we can guarantee that pin 5 will read 3 V by default when the button *is not* pressed. Alternatively, we can then expect a 0 V, or LOW signal, when the button *is* pressed.

7.4 DIGITIZING VOLTAGE

Great! Now we understand the `setup()` portion of this sketch and why we use a pull-up resistor. Now that our sketch is properly setup and configured, let's spend some time going over the contents of the `loop()` function.

```
//...
void loop(){
        // read the state of the pushbut ton value:
        buttonState = digitalRead(buttonPin);
//...
```

Here in `loop()`, we are introduced to a new function called `digitalRead()`. This function is unique and can be thought of as the counterpart to our `digitalWrite()` function we used in a previous chapter. `digitalRead()` is a prewritten function found in Energia that allows us to read the state of an input pin. Just like some of our other functions, we pass in a parameter to instruct the function which pin it should read. In this case, we want to read the input signal of pin 5, which is where our button is connected to. Depending on if the button is pressed or not, we can expect that pin to be either 3 V or 0 V, which correlates to a HIGH or LOW signal. The cutoff point between a HIGH or a LOW signal being read by a pin is actually at the halfway point between GND (0 V) and VCC (3 V).

This `digitalRead()` function also introduces a new concept in regards to functions. We see that this function actually generates something and produces a result. We are able to capture this result by storing it in a variable, which we called "buttonState." When this line of code is executed, the `digitalRead()` function reads whether pin 5 is a "1" or a "0" and spits it out. Lucky for us, our buttonState variable is there ready to catch our reading! We can see that we created this variable at the top of our code before the `setup()` function and initialized it to have a value of "0."

Now that *buttonState* holds the state of the button, "0" for pressed or "1" for not pressed, we then use some intelligence to make a decision based on that new information. We can see that if the button is pressed, buttonState = 0. Alternatively, if the button is not pressed, buttonState = 1. Take a look at the if-statement in the code, what do you think happens to the LED when the button is pressed versus not pressed? Does this match what you are seeing on your LaunchPad?

Depending on the state of the button, we set the green LED as either HIGH or LOW using the `digitalWrite()` function that we learned about earlier in this book. We notice that when the button is not being pressed, the green LED is turned on. Alternatively, when the button is pressed, the LED turns off. This might sound backward. Well, in a way it is.

The pull-up resistor that we are using causes the default state of the button to be a HIGH signal. This "backward" signal that we are reading from the button is what we call an "Active Low" signal. This is a shorter way of explaining that when the button is activated, the signal read by the pin is low.

As you can see below, the example sketch we use in this project leverages the simple if-statement that we introduced in an earlier chapter. The green LED reflects the electrical state of the output of the active low signal that our button generates.

```
//...

// check if the pushbutton is pressed. If button is pressed,
// buttonState = 0 (LOW)
if (buttonState == HIGH) { // button is not pressed
        digitalWrite(ledPin, HIGH); // turn LED on
}
else { // button is pressed
        digitalWrite(ledPin, LOW); // turn LED off
}
//...
```

Challenge:

Now that you understand the code, we challenge you to try to modify the code so that we can correct the inverted logic. What needs to happen to make the LED turn on when the button is pressed and off when the button is released?

7.5 BEYOND LEDS AND BUTTONS

Now that we have played with simple input and output signals, let's take a step back and think about what this *really* means. As exciting as toggling an LED might be, the concept of inputs and outputs is even more interesting! What a button (input) or an LED (output) is in our existing example could be easily replaced by a smoke detector (input) and a buzzer alarm (output). Or a rotary dial (input) could modify the speed of a spinning motor (output).

This is where microcontrollers get interesting and congratulations for making it this far! As we hinted at the beginning of this book, microcontrollers are tiny things that control stuff. With your knowledge of inputs and outputs, what types of things can you take control of, based on some logic and information collected from the microcontroller inputs? Think about the dream application you wrote down at the beginning of the book, then take the time to write down the inputs and outputs you think you need to accomplish your dream project in the space below.

My dream microcontroller application will have:

Inputs (senses): _____

Outputs (actions): _____

If you think you have the right tools to implement some of the inputs and outputs above based on what we have learned about so far, put a checkmark next to it. We should now be able to read in digital inputs and generate a digital output using the `digitalRead()` and `digitalWrite()` functions. If you have some inputs and outputs that are still unchecked in your list above, keep them in mind as we continue our journey!

7.6 PROJECT 6: TURN ANYTHING INTO AN INPUT

As we learned in this chapter, our LaunchPad brain can read and write in 1s and 0s, or HIGH and LOW signals. We've also learned that these HIGH and LOW signals are a representation of the presence or absence of a voltage. When an input pin reads a 3-V signal using the `digitalRead()` function, it causes that port pin to see a HIGH signal, or a 1. When the input pin reads 0 V, it causes that port pin to see a LOW signal, or a 0.

In the previous project, we used a pushbutton to control the flow of electricity to control the signal being read by port pin 5. In this project, we'll use different types of objects to control the flow of electricity by turning various items into buttons!

But first, let's set up our hardware. Here are the ingredients.

Ingredients:

1. LaunchPad
2. A tablespoon (not of anything, just a metallic tablespoon)
3. 2 × Dual-ended alligator clip wires
4. Educational BoosterPack
   OR
   Breadboard and 22 MΩ resistor

If you are using the Educational BoosterPack, plug it into your LaunchPad. On the Educational BoosterPack, you'll notice that we have four holes in the middle of the BoosterPack. These four holes are grouped into pairs. One pair is labelled as P2.7, while the other pair of holes is labelled as GND. These holes were designed to make it very easy to clip on a pair of alligator clips.

For those that do not have an Educational BoosterPack yet, here is the schematic of how we can set it up using a breadboard and a few components.

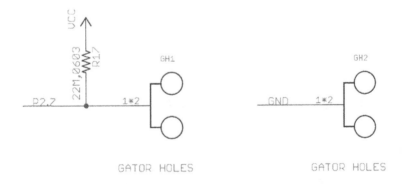

Here's a more friendly version of the schematic made with a program called Fritzing.

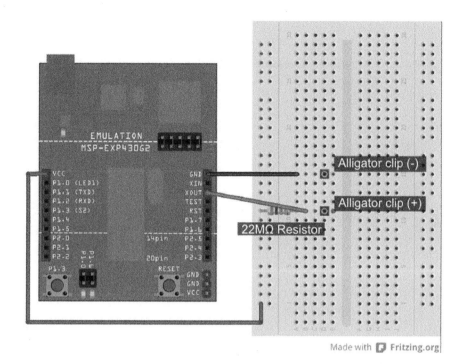

Once we have the hardware set up, we will be using two dual-ended alligator clip wires to connect our LaunchPad pins P2.7 (labelled as XOUT on the LaunchPad) and GND to the outside world! This setup will allow us to turn anything that is conductive into a button.

But first, let's get the software up and running inside of Energia. As we're essentially creating a button, we can use the same Button code example that we played with earlier in this chapter. However, there are some slight modifications that must be made and are shown on the next page.

```
const int buttonPin = 18; // the pin number that our alligator clip is
                          // connected to (P2.7)
const int ledPin = GREEN_LED; // the number of the LED pin

// variables will change:
int buttonState = 0; // variable for reading the pushbutton status

void setup() {
        // initialize the LED pin as an output:
        pinMode(ledPin, OUTPUT);
        // initialize the pushbutton pin as an input:
        pinMode(buttonPin, INPUT);
}

void loop(){
        // read the state of the pushbutton value:
        buttonState = digitalRead(buttonPin);

        // check if the pushbutton is pressed. if it is, the buttonState
        // is LOW
        if (buttonState == HIGH) {
                // turn LED on:
                digitalWrite(ledPin, HIGH);

        }
        else {
                // turn LED off:
                digitalWrite(ledPin, LOW);
        }
}
```

First, we need to change the buttonPin assignment to the port pin that our alligator clips are attached to. In this case, we have the alligator clips connected to P2.7 (pin 18).

Second, we need to disable the internal pull-up resistor in our LaunchPad. We do this by passing the following parameters into our function, pinMode(buttonPin, INPUT). The reason we simply use "INPUT" instead of "INPUT_PULLUP" as we did in the previous example is because our Educational BoosterPack (as well as our DIY breadboard design) utilizes an external pull-up resistor of 22 MΩ.

But, why don't we simply use the internal pull-up resistor inside of our LaunchPad? Good question. It's all about voltage! To explain this, we have to understand how the value of our pull-up resistor can affect the flow of electricity. The pull-up resistor we need in this project (22 MΩ) is much larger compared to the internal resistor of the LaunchPad (10 kΩ). Well, before we go into details here, let's first do something fun with our hardware and code! Compile and download the code into the LaunchPad.

Step 0: Ensure your hardware is set up either using the Educational BoosterPack connected to a LaunchPad or rig up your breadboard design based on the diagrams shown earlier. Then, compile, download, and run the modified "button" code example inside of Energia.

Step 1: To start things off, let's use a metallic spoon. Clip one end of the alligator clip to your spoon, then clip the other end of the wire to the holes associated to P2.7.

Step 2: Next, take your second wire and clip one end to the holes associated to ground (GND). Now, hold on to the other end of the grounded wire in your left hand.

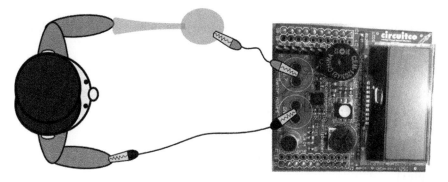

Step 3: While holding on to the grounded wire in one hand, touch your button (spoon) with your free hand. At that point, your green LED should turn off! Success, you turned something into a button!

7.6.1 What's Going on Here?

By holding onto the grounded wire, the instant we touch our spoon button, we close a circuit, similar to the pushbutton in the previous example. When we close the circuit by touching the button, we cause our port pin P2.7 to read a LOW signal (remember, it's an active low signal due to our pull-up resistor). Once port P2.7 reads a LOW signal, our code knows a button has been pressed and tells the LED to turn off.

Fun! So play around with different items to see if they are conductive enough to become a button for your LaunchPad! This should open up thousands of possibilities!

Here are some ideas.

Rain detector [input = water/output = LED]

Have both the P2.7 and the GND probes exposed in a cup. When the cup fills with a certain amount of rain water, the water will close the circuit between P2.7 and GND. When this circuit closes, we can illuminate an LED to tell you that it's raining.

Or a twist on this project could be a reminder for watering your plants or refilling your pets water bowl. When the plant or pet has water to drink, the circuit is closed. Once the water starts to dry up and reach a certain level, below our alligator clip probes, the circuit is no longer closed and an LED can glow to remind us to add more water.

Alarm system to prevent (barefoot) bandits [input = doorknob/output = buzzer]

Connect P2.7 to a metallic doorknob. Connect the GND probe to a sheet of aluminum foil on the doormat. When someone (barefoot) stands on the aluminum foil and touches the doorknob, it closes the circuit and causes a buzzer to buzz!

Hi-Five counter [input = your friend/output = increment counter on a display]

Hold on to the GND clip in one hand. Instruct a friend of yours to hold on to the P2.7 clip in one of their hands. Then, using your free hands, hi-five each other! The hi-five will close the circuit and the LaunchPad will increment a number on an LCD screen to keep count of your hi-fives.

Have some fun with this concept. Play around with different types of inputs to cause your LED to blink. Our alligator clips help us expand on the concept of inputs and outputs. As you clip your LaunchPad to different items, you'll be pretty amazed at how many different types of applications you can create using the same basic code and hardware!

Now that we've had some fun with our alligator clips, let's go back to a question we had earlier. Why are we using a massive 22 MΩ resistor as opposed to the 10 kΩ resistor that is available inside of our

LaunchPad? As we introduced earlier, the answer is presented to us as we understand how voltage is affected by resistors.

7.7 EE101—OHM'S LAW

A *resistor* is a basic component found in many electronics, and it restricts the flow of electricity, which we call current. We can relate the three major components of electricity using a relationship that is known as *Ohm's Law*. Ohm's Law simply shows the relationship between resistance, current, and voltage and is shown as followed:

$$\text{Voltage} = \text{current} \times \text{resistance, or } V = I \times R$$

Voltage (V) is the electrical potential measured between two points. It is measured in Volts (V).
Current (I) is the measurement of electrical flow. It is measured in Amps (A).
Resistance (R) is the measurement of how restrictive a material is to the flow of electricity. It is measured in Ohms (Ω).

We can take a few things away from this simple equation. Let's see how these three elements are related to each other.

$$V = I \times R$$

We can see that voltage, or the electrical potential between two points, is directly proportional to I and R. This means that if current or resistance increases, the electrical potential between two points will also increase.

$$I = V/R$$

This shows us that current is inversely proportional to resistance. This means that if resistance increases, current decreases. This makes sense because resistance is the restriction of electrical flow, or current.

To further explain this concept, let's see how this all relates to how fluid flows through pipes. Like water flowing through pipes, Ohm's Law helps us understand how electricity flows through a circuit.

Electrical current is similar to the flow of a fluid. The volume of fluid that flows through a particular section of "pipe" is similar to the amount of current, or electricity that flows inside of a circuit or wire.

Similarly, resistance of a material measures how "restrictive" the material is to the flow of electricity through that particular material. This is similar to the diameter of our pipe in our water flow analogy. The smaller the diameter is of our pipe, the more restrictive, or resistive, it will be to the flow of water.

Lastly, voltage, which is the measure of electrical potential between two parts of a circuit, is similar to the difference of fluid pressure at two different parts of our pipe system.

Let's look at a simple example.

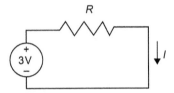

In this example circuit, we have a battery that creates the electrical potential, or voltage (V), and a resistor (R) that restricts the flow of current (I). As you can see in this example, current always flows in the direction from a high electrical potential to a low electrical potential (flows from the positive terminal to the negative terminal of the battery). Also, it is important to note that in this setup, current flow is constant and equal throughout the entire circuit.

In the circuit above, we chose a 3-V battery. In the physical world, we can use 2 AA batteries to create a ~3-V power source for many LaunchPad-based projects, as each AA battery can typically source about 1.5 V. Alternatively, when plugged into your PC over USB, a stable 3.6 V power source is made available for your LaunchPad.

The power supply pin for our LaunchPad is referred to as VCC. When we connect the VCC pin to a power supply (either from external batteries or via USB), our LaunchPad is powered and can start executing code! Our LaunchPad brain is capable of running off a voltage as long as it's within the range of 1.8–3.6 V. However, in this section of our book, we'll be using a 3-V battery source.

If we had a voltage measurement tool, also called a voltmeter, we would read a voltage of 3 V on the positive end of the battery. On the negative terminal of the battery, our voltmeter would read 0 V. What caused this voltage drop between the positive and the negative terminal of the battery?

The answer is the resistor! Our resistor, R, caused the 3 V voltage drop between the positive and the negative terminal of our battery. It is also important to note that the voltage measured at the positive terminal of the battery (3 V) is exactly the same as the reading our voltmeter would give us if we measured just to the left of the resistor. Similarly, we would get a 0-V reading at the negative end of our battery as well as a 0-V reading at the point just to the right of our resistor. The voltage is consistent at any point on the wire because the wire has virtually no resistance. With Ohm's Law, we can find the current flowing through our circuit as well as verify the voltage drop across our resistor.

Now that we have a basic understanding of Ohm's Law, let's look at a simple example. In the circuit below, how much current is flowing through our circuit?

Remember, Ohm's Law is $V = I \times R$

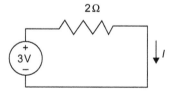

Based on Ohm's Law, we know that we can find current using the following relationship:

$$I = V/R.$$
$$I = 3\ V/2\ \Omega = 1.5\ A$$

Awesome! The current that is flowing through the system above is 1.5 A. Now that we know what the current flowing through our circuit is, we can find out what the voltage drop across our resistor is.

Voltage drop across the 2 Ω resistor can be found using Ohm's Law once again! We can use the relationship $V = I \times R$.

$$V_{\text{drop across 2 }\Omega} = 1.5 \text{ A} \times 2 \text{ }\Omega = 3 \text{ V}$$

So the voltage drop across the 2-Ω resistor is 3 V, which is what we expected.

Let's look at something a little more complex. What if we introduce another resistor into the circuit? How will the flow of electricity be affected?

Much like our original setup with just one resistor, current is consistent and equal throughout the entire circuit. Also like our original example with just one resistor, we know that there is a voltage drop across each resistor, $R1$ and $R2$. In this setup, we call the resistors to be set up in "series" with each other. When a resistor is in series with each other, we can find the total resistance by simply adding them together ($R1 + R2$), so they operate like one big resistor. So let's use this knowledge and apply it to Ohm's Law:

$$V = I \times (R1 + R2)$$

Or, another way of looking at is this:

$$V = (I \times R1) + (I \times R2)$$

So what we end up seeing is the voltage drop associated to each resistor!

- $V_{\text{drop across } R1} = I \times R1$

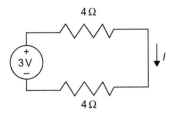

In this setup, we call the resistors to be placed in "series" with each other. When a resistor is in series with each other, we can find the total resistance by simply adding them together ($R1 + R2$), so they operate like one big resistor. In our case, the total resistance of the two resistors is $2 \text{ }\Omega + 4 \text{ }\Omega = 6 \text{ }\Omega$. With the help of Ohm's Law, we also know that the current throughout this entire circuit is equal to the voltage divided by the total resistance ($I = V/R$).

- Current = 3 V/(2 Ω + 4 Ω) = 0.5 A

Ohm's Law also tells us what the voltage drop is across each resistor:

- $V_{\text{drop across 2 }\Omega}$ = 2 Ω × 0.5 A = 1 V
- $V_{\text{drop across 4 }\Omega}$ = 4 Ω × 0.5 A = 2 V

This makes sense since the total voltage drop across the two resistors in series should be 3 V. We can also see that the voltage drop is proportionately distributed between the two resistors based on their resistive value. This is what we call a *voltage divider*. With the two resistors, we "divide" or split the total voltage drop between the two resistors in a proportional manner based on the individual resistances.

So let's pull out our imaginary voltmeter to see what kinds of measurements we should see at different points of this two-resistor circuit. Let's follow the flow of current in our system starting at the positive 3 V terminal of our battery.

- As we would expect, the voltage at any place between the positive terminal of the battery and the 2 Ω resistor is 3 V.
- Also, just like our single-resistor example, the voltage anywhere between the negative terminal of the battery and the 4 Ω resistor is 0 V.
- But what about the voltage reading between the 4 Ω and the 2 Ω resistor? Well, that's where understanding the individual voltage drops come into play.
  - The voltage measurement that would be read between the 2 Ω and the 4 Ω resistor is 2 V. The reason for that is because the 2 Ω resistor causes the voltage to drop from 3 V down to 2 V due to the 2 Ω resistor's 1 V voltage drop.
  - Finally, the 4 Ω resistor causes the final voltage drop from 2 V down to 0 V. This is perfect because we know that the measurement on the other side of the 4 Ω resistor is 0 V because it is connected to the negative terminal of the battery.

Congratulations! You just took your first step into the world of electrical engineering! Hope you enjoyed your stay!

7.8 YOUR *RESISTANCE* IS FUTILE

Now that we are a few concepts closer to an electrical engineering degree, let's see how this all applies to our alligator clip project and why we need that huge 22 MΩ resistor! First, let's look at the hardware setup of our alligator clip project. In the example circuit below, we are using a spoon as our "button" or input.

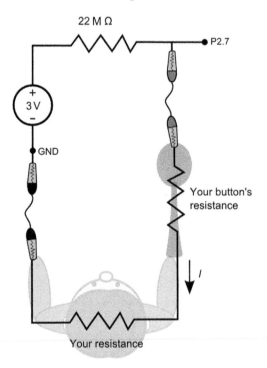

As you can see, this setup has three resistors in series. We have our pull-up resistor, we have our spoon/button, and we also have you!

We know that the LaunchPad is watching our input pin 18 (P2.7) according to our code. By default, P2.7 reads a HIGH signal because our pull-up resistor connects P2.7 to the positive terminal of the 3 V battery by default. In order for P2.7 to read a LOW signal, we need P2.7 to read a voltage that is closer to 0 V. We do that by closing the circuit by holding on to the alligator clip connected to ground (0 V) with one hand and touching the button (in this case a spoon) with other hand. By closing the circuit, we create a voltage divider. In order for the LaunchPad to see the right voltage at P2.7 to read a LOW signal, we need:

(Voltage drop across you + button) < voltage drop across the
22 *MΩ* resistor

The reason we need a huge 22 MΩ resistor is because of the voltage divider that we created. We need to ensure that when we close the circuit by holding onto the ground wire and the spoon, the voltage that is seen by P2.7 is less than 1.5 V (VCC/2). We need to ensure P2.7 sees a voltage less than 1.5 V because that is the threshold as to which our input pin P2.7 will see either a HIGH or a LOW signal.

This means that as long as the combined resistance of your body and button ($R_{you} + R_{button}$) is less than 22 MΩ, your circuit should work! So to make this particular application works, we want our pull-up resistor to be very large to account for the combined resistance of you and the button. If we used the internal 10 kΩ pull-up resistor that is available inside of our LaunchPad, we would not be able to make the voltage read by P2.7 be below 1.5 V because the resistance of a typical person is much larger than 10 kΩ.

7.8.1 Ohm's Law and the Digital World

To summarize, Ohm's Law helps us understand how circuits can manipulate the flow of electricity. With some clever circuit creation, we can translate electrical signals into HIGH or LOW digital signals that our LaunchPad can recognize. These HIGH and LOW signals that are received by the pins of our LaunchPad are received from the outside circuitry and seen as inputs. Based on these inputs our LaunchPad can leverage its intelligence to create interesting and useful outputs. Thanks for sticking with us as we explored through the world of digital inputs and outputs.

CHAPTER 8

Analog: The Infinite Shades of Gray

8.1 BEYOND BLACK AND WHITE

So far, we have learned that microcontrollers think in a digital or binary language of 1s and 0s. In the previous chapter, we learned how digital I/O can unveil a thin layer of external interaction, that allows the LaunchPad to see the world in black and white, experiencing just a single bit at a time. In this chapter, we will take it to the next level and enable the LaunchPad to experience the full spectrum and interact with HIGH and LOW signals and everything in between!

An analog signal is a voltage that may fluctuate anywhere between 0 V and VCC (~3 V). While all of the input pins on our LaunchPad can differentiate between 0 V and VCC and round them up or down to a digital value of 1 or 0, only some pins on our LaunchPad can read the full range. It is through these specialized analog input pins where we can get a better view of the outside world.

8.2 PROJECT 7: OUR FIRST ANALOG EXPERIMENT— THE POTENTIOMETER

Let's play with our first analog sketch. In this first project, we will interface with our first analog component, the analog potentiometer. Essentially, a potentiometer is a specialized resistor that can have a variable resistance. On a manually controlled potentiometer, which is what we are using today, a rotational dial can be used to vary the resistance from 0 to the maximum resistance value, which is 10,000 Ω for our project. As we rotate the dial and change the resistance, we alter the output voltage that comes out of the potentiometer. We will then use one of the special analog input pins on our LaunchPad to read the analog signal coming out of the potentiometer.

Note

The potentiometer actually works using the same voltage divider principle that we learned about in Chapter 7. Think of the potentiometer as being made up of two resistors ($R1$ and $R2$) whose individual resistances add up to be the total resistance of the potentiometer. As you turn the dial, you are essentially changing the distribution of resistance across these two resistors. By changing the distribution of voltage between $R1$ and $R2$, we can change the voltage that is read at the point between the two resistors, which is the output of our potentiometer. The image below should give a better understanding of what's happening inside the potentiometer.

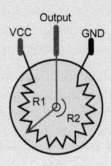

Ingredients:

1. LaunchPad
2. Educational BoosterPack
   OR
   Breadboard and 10,000 Ω potentiometer

Let's get the hardware set up using the steps below.

1. First, position the RXD and TXD jumpers as shown below (this is for LaunchPad rev 1.4 and newer). If you are using an older revision of the LaunchPad, remove the jumpers RXD and TXD and cross them using jumper cables.

Ver 1.3 and older: Jumper positions

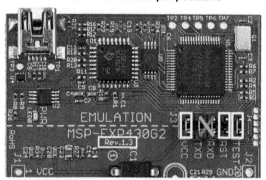

Ver 1.4 and newer: Jumper positions

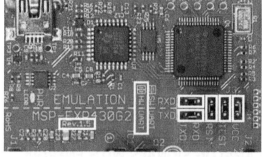

2. If you have your Educational BoosterPack handy, simply plug the BoosterPack onto the LaunchPad. The Educational BoosterPack has a potentiometer on the board. The potentiometer on the BoosterPack is connected to pin 5 on our LaunchPad, which just so happens to be an analog input channel—channel A3 to be exact.

3. On the Educational BoosterPack, make sure that the jumper J5 is connected in the A3_POT position. This will ensure that our potentiometer is connected to our A3 input pin.

Place Jumper J5 in the A3_POT position

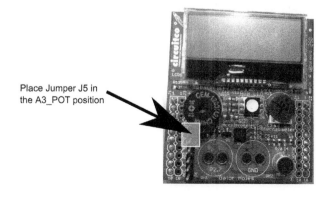

If you do not have an Educational BoosterPack handy, make sure to wire up your potentiometer to your LaunchPad as shown in the diagram below.

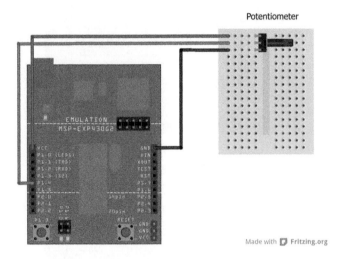

Alright, now that our hardware is all set up, plug your LaunchPad into your computer and let's get into the software.

8.3 READING ANALOG SIGNALS WITH ENERGIA

Fire up Energia. In the new empty sketch, let's set up the framework of our Energia sketch: `setup()` and `loop()`.

Nothing needs to be configured in the `setup()` function at this point, so let's keep it empty and skip to the `loop()` function.

From the schematic above, we can see that the potentiometer output pin is connected to pin 5, or analog channel 3 (A3). We know that this is analog channel A3 thanks to that nifty pin assignment image that is available in the Appendix: *Quick References* section of this book. In order to read the analog value that is seen at this pin, we need the following lines of code in our `loop()` function.

```
setup(){    // we don't have to configure analog pins as input or output
            // to use them
}

loop(){
    int analogValue;    // create a variable to store the analog reading
    analogValue = analogRead(A3);
}
```

The first thing we needed to do was create a variable that will catch our analog reading.

Secondly, we used a new function called `analogRead()`. This is very similar to our `digitalRead()` function in that it also returns a value. As you know, `digitalRead()` returns a value of 0 or 1. In the case of `analogRead()`, we can expect an output value between 0 and 1023, where 0 and 1023 are mapped linearly to a voltage reading between 0 V and VCC (~3 V). So, if we have a reading of ~1.5 V (VCC/2), we should get an output of 512 from our `analogRead()` function.

OK cool, we have acquired our first analog reading! But, now what? Well...we know that our LaunchPad is reading a number anywhere between 0 and 1023 depending on the position of the potentiometer and storing it in the variable *analogValue*. As *analogValue* is a number, we can compare it against another value using an if-else-statement to do different things depending on the reading.

Let's go ahead and add the following lines to your `loop()` function.

```
setup(){ // we don't have to configure analog pins as input or output to
         // use them
}
loop(){
        int analogValue; // create a variable to store the analog
                         // reading
        analogValue = analogRead(A3);
        // this is the halfway point between 0 and 1023, the range of
        // our analog input reading
        if (analogValue > 512){
                digitalWrite(ledPin, HIGH); // turn LED on
        }
        else{
                digitalWrite(ledPin, LOW); // turn LED off
        }
}
```

The if-else-statement we are using in the code turns the LED ON or OFF depending on the analog reading that is generated by the potentiometer. If *analogValue* is greater than 512, the LED is ON. Otherwise, the LED is OFF.

Alright, we're almost ready to fire up our new sketch on the LaunchPad, we just need to add the configuration code for the LED that will turn ON or OFF depending on the position of the potentiometer. If you are using the Educational BoosterPack, you can use the white LED on the BoosterPack, which is located at pin 13.

Insert this line at the beginning of the sketch:

```
int ledPin = 13; // white LED on Educational BoosterPack
```

If you are breadboarding your own Potentiometer circuit, you can reuse one of the LaunchPad onboard LEDs, such as the RED_LED located at pin 2. We can do this by adding this line at the beginning of the sketch:

```
int ledPin = 2; // Red LED on the LaunchPad
```

Now that we have set the appropriate number using the ledPin variable, let's add the following code to configure the LED pin as an OUTPUT in the **setup**() function:

```
int ledPin = 2; // Red LED on the LaunchPad
// int ledPin = 13; // white LED on Educational BoosterPack
// comment out one of the lines above depending on which LED you're
// using

void setup(){
        // we don't have to configure analog pins as input or output to
        // use them
        pinMode(ledPin, OUTPUT);
}

void loop(){
        int analogValue; // create a variable to store the analog
                         // reading
        analogValue = analogRead(A3);
        // this is the halfway point between 0 and 1023, the range of
        // our analog input reading
        if (analogValue > 512){
                digitalWrite(ledPin, HIGH); // turn LED on
        }
        else{
                digitalWrite(ledPin, LOW); // turn LED off
        }
}
```

And... we're done! Go ahead and click Verify and Download on Energia to load this on the LaunchPad.

Once the code is up and running on the LaunchPad, take note of the LED (white LED on the Educational BoosterPack or the red LED on the LaunchPad) and turn the potentiometer. If we did everything correctly, we can get an idea of what the analogValue is based on the status of the LED. If the LED is on, that means the `analogRead()` function returned a value greater than 512. If the LED is off, our `analogRead()` function returned a value less than or equal to 512.

Start turning the potentiometer in either direction until the status of the LED changes (on→off or off→on) or until you can no longer turn the potentiometer. Spend a few minutes playing with it, can you eventually find out at which position of the potentiometer the LED toggles?

Alright, we completed our first analog project! Some may think, "Gee great! We've learned all of this analog stuff just to create a glorified button to turn an LED on and off. I could have used a digital one for that."

Well, patience, eager one! Let's go back to our toolbox and see what else we can use to interpret the values being returned by our `analogRead()` function. Let's try to use the `delay()` function to visually demonstrate the analog readings. If you remember, `delay(x)` will cause our LaunchPad to pause and take a breather for x milliseconds. Let's combine these two functions and see if the sense of time can help us represent the incoming voltage readings.

8.4 PROJECT 8: ANALOG-TO-DIGITAL SIGNAL CHAIN

In this project, we are going to use two functions that we have learned about, `analogRead()` and `delay()`. With the help of these functions, we are going to create an analog-to-digital signal chain.

Define

Signal Chain

Signal chain is an electrical term that describes a series of signal acquisition, conditioning, and output. In this tandem series, the signals are "chained" together so that the output of the previous stage is the input for the next stage. A signal chain can be used to gather data, process said data, and output based on the refined data.

In this case, we are creating a signal chain that has an analog input and a digital output. We will read the analog reading coming out of the potentiometer and will use that data to change the speed of a blinking LED (digital).

To get this project started, we'll modify the same code that we wrote in the previous laboratory. Replace the lines of code inside of the loop() function with the following:

```
...
void loop(){
        int analogValue; // a variable to store the analog reading
        analogValue = analogRead(A3); // read analog output of
                                      // potentiometer

        // turn the ledPin on
        digitalWrite(ledPin, HIGH);

        // stop the program for <analogValue> milliseconds:
        delay(analogValue );

        // turn the ledPin off:
        digitalWrite(ledPin, LOW);

        // stop the program for for <analogValue> milliseconds:
        delay(analogValue );
}
```

Once you have replaced the loop() function contents with the new code, take a couple of minutes to digest the code to figure out what it does. Then verify your conclusion by downloading the code to the LaunchPad and see if your hypothesis was correct.

You should see that in this project, the LED blinks faster or slower depending on the position of the potentiometer. That is because the varying analog voltage reading that the potentiometer outputs becomes the input to the digital delay() function. This helps us translate the analog reading into something that is visible to us.

We now have a better idea of the analogRead() function than what we originally had with our simple LED ON/OFF example from the previous project. Congratulations, you have successfully created your first analog-to-digital signal chain!

By the way, what you have created so far is a recreation of an existing example sketch provided in Energia. You can find it by going to *File > Examples > 3.Analog > AnalogInput.*

So far we have found a couple of ways of interpreting the analog signal. But we still have not been able to truly receive or know the exact values we get back from the potentiometer. This is mainly due to the fact that we haven't learned a way to get the information from the LaunchPad back to us yet.

8.5 PROJECT 9: OUR FIRST DEBUG SESSION—A LOOK INSIDE OUR LAUNCHPAD BRAIN

When creating or running a new Energia sketch on your LaunchPad, sometimes we will want to know the status of the LaunchPad or see certain elements inside of your sketch. Up until this point, we have been relying on our trusted LEDs to let us know what's going on inside the vast and mysterious brain of the LaunchPad. A few LEDs unfortunately are quite limited in terms of information that can be transmitted and are a limited way of understanding how our LaunchPad is doing. In more complicated projects, there will be a need for a better way of diagnosing or investigating what's going on inside of our LaunchPad. Enter debug.

Define

Debug
   The process of identifying and removing errors from computer hardware or software.

To help developers check on the status of the LaunchPad or diagnose an issue within a sketch, Energia offers a way for the LaunchPad to send messages back to Energia via an included Serial Monitor. So let's give this Serial Monitor a try and see how we can report back the values of analogRead() to the computer.

To enable our LaunchPad to send diagnostic data back to the computer, we need to initialize it using the following line of code in our setup() function.

```
void setup() {
        //...
        Serial.begin(9600); // initialize capability to send data back
                            // to Energia
        //...
}
```

The number 9600 passed in as a parameter for the `Serial.begin()` function identifies the speed at which the LaunchPad communicates back to the computer. In this case, 9600 means 9600 baud. We'll learn more about this serial communication capability in Chapter 10. However, for the time being, let's simply use it for our debugging purposes.

Once configured in `setup()`, this new capability of sending diagnostic information back to the computer is ready for us to use in the main `loop()` function.

Every time you want to send a message back to the PC, simply use the following new functions:

If you want to send a fixed text message, or string:

```
Serial.print("Your potentiometer reading is = ");
```

Or, if you want to send an integer value that is stored in a variable:

```
Serial.println(analogValue, DEC);
```

The DEC in the second example instructs our function to display the value to be printed in a decimal format. As you can see, these functions are very useful because we can pass in a variable that is stored inside of our LaunchPad brain and print out the value that is stored in the variable on the computer screen. This, in short, is the essence of debugging inside of Energia. We can strategically place these print functions inside of our code to send out diagnostic data back to the computer to help us understand what is happening inside of our LaunchPad brain.

Another thing to note is that we are using two types of print functions:

```
Serial.print(); // prints out whatever you pass in as a parameter
```

```
Serial.println(); // Same as above, except this also includes a linefeed
                   // after it prints
```

Let's use these new functions to take a peek inside of our LaunchPad brain and print out the analog readings that are received from the potentiometer.

Let's go back to our bare-bone potentiometer sketch in Energia that we used in Project 7. Let's initialize this new debug capability in the `setup()` function and simply insert a few print functions at the end of our `loop()` function to read out the *analogValue*. We should get the following.

```
int ledPin = 2; // Red LED on the LaunchPad
// int ledPin = 13; // white LED on Educational BoosterPack
// comment out one of the lines above depending on which LED you're
// using

void setup(){
        // we don't have to configure analog pins as input or output to
        // use them
        pinMode(ledPin, OUTPUT);
        Serial.begin(9600); // initialize capability to send data back
                            // to Energia
}

void loop(){
        int analogValue; // create a variable to store the analog
                         // reading
        analogValue = analogRead(A3);
        /* this is the halfway point between 0 and 1023, the range of
        our analog input reading*/
        if (analogValue > 512){
                digitalWrite(ledPin, HIGH); // turn LED on
        }
        else{
                digitalWrite(ledPin, LOW); // turn LED off
        }

        Serial.print("Your potentiometer reading is = ");
        Serial.println(analogValue, DEC);
}
```

Let's make sure our code checks out, so hit the Verify and Download button in Energia to download it into our LaunchPad.

Once your LaunchPad has been reprogrammed and starts running, we know that it is sending messages back to your computer thanks to the new lines of code that we have added to our sketch. In order to receive those messages, let's click on the magnifying glass that is on the top-right corner of the Energia window. This button is labelled as the "Serial Monitor."

This Serial Monitor is a nifty tool inside of Energia, that displays any messages coming from our LaunchPad.

If you see an error when trying to open your Serial Monitor, make sure you have the correct serial port selected in the "Tools" menu inside of Energia. Be sure to select the right port for your setup. While the screenshot shows COM19, your communication port number may be different. In most cases, the one you want to use will be the last one on that drop-down list.

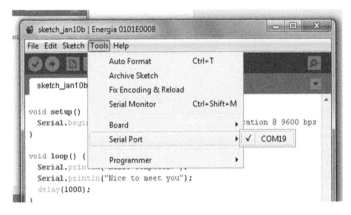

Once you open up the Serial Monitor window, we should start to see a stream of data flowing into our computer screen.

What we're seeing is the output of our `analogRead()` function. Turn the potentiometer wheel. What happens?

We should start to see the incoming values vary based on the position of our potentiometer wheel.

As you can see the `analogRead()` function returns an integer value between 0 and 1023, which somehow represents the voltage level present at the pin. When we turn the wheel all the way to one direction, we should get a value of 0 on the Serial Monitor. When the wheel is turned completely in the other direction, we should see a value of 1023. With the potentiometer, we can actually see the full range that our `analogRead()` function is capable of reading.

8.6 DIGITIZING AN ANALOG SIGNAL

8.6.1 Analog-to-Digital Conversion (ADC)

An analog value is similar to a real number in the sense that it is usually not a round, whole number, but instead a value of infinite precision (x.yyyyyyyyyyy...). This infinite precision unfortunately can't be completely and fully captured by any instrumentation. In reality, our LaunchPad's digitally limited brain can only measure and approximate the value to the best precision it is equipped with.

The process of converting an analog reading into a digital (integer) value is called analog-to-digital conversion (ADC or A/D). The ADC instrumentation has a finite number of digital steps that it can map from the infinite range of an analog signal. The *resolution* of the converter indicates the number of unique readings it can produce over the entire analog range. The values are usually stored digitally in binary form, so the resolution is usually expressed in bits. In consequence, the number of discrete values that an ADC can read is a power of 2.

Define

Resolution
Determines the number of unique readings an ADC module can recognize.

For example, an ADC with a resolution of 10 bits (like the one on our LaunchPad) can convert an analog input and read it as a value between 0 and 1023. This means we can differentiate 1024 unique values with the LaunchPad's ADC module. So, why 1024?

It all comes back to binary values. Because the ADC module in our LaunchPad has a *resolution* of 10 bits, this means the largest value we can represent is 1111111111_2 in binary. And using the skills we learned in Chapter 6, we know that this is equal to 1023_{10} in decimal.

8.6.2 How Do I Know What Voltage Level My `analogRead()` Value Is Correlated to?

As we know the resolution of our LaunchPad's ADC is 10 bits, we know that the ADC can map a range of voltage levels to a reading in the range of $0-1023$.

The voltage range that is detectable by the ADC is determined by the voltage reference that the ADC is configured to. By default, the voltage reference of the LaunchPad ADC is configured to VCC (~ 3 V). This means that the LaunchPad ADC's recognizable voltage range is between 0 V and VCC.

For example, if the LaunchPad is connected over USB, the source voltage (VCC) is 3.6 V. This means that the ADC can read in an analog voltage range of $0-3.6$ V. The `analogRead()` function then maps this analog reading to a value in the range of $0-1023$. Hence there are 1024 voltage steps, each with the width of 3.6 V/1024 $\cong$ 0.00352 V = 3.52 mV, that are distinguishable by the ADC.

The 3.52 mV step size is the smallest amount of voltage that can be detected by the ADC based on its voltage reference and resolution. Therefore, if the voltage level being seen at a pin is not actually equal to one of the 1024 steps (or a multiple of 3.52 mV), it will be rounded up or down to the closest step. Because of this rounding, we can begin to understand the importance of resolution and the voltage reference that is used and how it affects the precision of our analog readings.

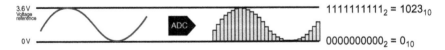

8.7 PROJECT 10: BATTERY-LIFE TESTER

At this point, we know how to read an analog reading using our `analogRead()` function. We also know how to take our digitized ADC reading ($0-1023$) and correlate that to an actual voltage level (0 V to reference voltage (typically 3.6 V)). In this project, we are going to use our LaunchPad's analog input pin to read the voltage of an AA battery. AA batteries, when fully charged, usually have a voltage level of around $1.5-1.6$ V. If we do the math, we can expect that at a voltage reading of ~ 1.5 V, we should expect our `analogRead()` function to return a mapped integer value of ~ 426.

Throughout the life of an AA battery, not only does the charge in the battery go down, but the voltage also starts to dip. Eventually, the battery's charge and voltage will drop all the way down to its end-of-life limit, which for an AA battery is typically around ~0.9 V.

At 0.9 V, we can expect our `analogRead()` function to return a value of ~256. So to check if a battery has any life left in it, we can use our `analogRead()` function to check if the raw ADC reading is >256. If not, then we know that this battery has no life left to be usable.

Ingredients:

1. LaunchPad
2. 1AA Battery
3. 2 Wires/probes

The hardware setup for this project is simple! We just need two wires, which are going to connect our LaunchPad pins to the two terminals of an AA battery. Using one of the analog input channels on our LaunchPad, we'll be able to read the voltage of the battery.

Connect one wire to pin 6 (P1.3), which is also the pin that is connected to analog input channel 4 (A4). Then, take the other wire and connect it to the GND pin on the LaunchPad.

In this project, we will be using the LaunchPad's onboard LEDs, so if you have taken off the LED jumpers in a previous project, you will need to place these jumpers back for this exercise.

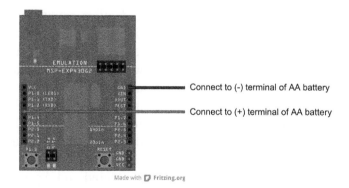

Made with Fritzing.org

Hardware is done! Before we start probing our battery, let's turn to Energia and write our sketch.

```
int sensorPin = A4;  // select the input pin for the battery
int ledPin = 14;     // select the pin for the LED (Green LED = 14)
                     // variable to store value coming from sensor

void setup() {
          // declare the ledPin as an OUTPUT:
          pinMode(ledPin, OUTPUT);
}

void loop() {
          // read the value from the sensor:
          sensorValue = analogRead(sensorPin);

          if(sensorValue > 256){
                    // turn the ledPin on because battery still has
                    // life left in it!
                    digitalWrite(ledPin, HIGH);
          }
          else{
                    // battery is dead, turn LED off
                    digitalWrite(ledPin, LOW);
          }
}
```

As you can probably deduce from the code, we are simply turning the green LED on our LaunchPad on if we still see that the voltage reading is greater than 0.9 V (which maps to ~256). If the battery appears to be dead, then we use our if-else-statement to turn the LED off. Don't forget to correctly recycle your dead batteries!

8.8 TYPES OF ANALOG SIGNALS

As mentioned before, the world is vast and offers countless of things that can be sensed, whether it be color, size, shape, humidity, pressure, etc. To allow our LaunchPad, or any microcontroller for that matter, to sense these wonderful flavors of the world, we need to first of all translate them into a language that it can understand. This is where electrical sensors come into play. The analog sensors can essentially convert all sorts of real-world "senses" into a voltage signal that we can then read with one of the LaunchPad analog pins. Here are a few examples of some of the exciting senses that can be translated into voltage levels, which can then be ultimately sensed and digested by the LaunchPad.

8.8.1 Temperature

The temperature sensor is probably one of the most popular and useful analog sensors in an embedded application. Essentially an analog temperature sensor converts the ambient temperature into a voltage reading, which can then be sensed by an analog pin of a microcontroller. The temperature sensor is great for applications that record and display temperature information such as a thermometer when you are sick, or the thermostat for your air conditioner. However, knowing the temperature also helps in applications where the microcontroller makes decisions based on temperature to maintain a larger system. In this case, the microcontroller may compensate components if it's too hot or cold, or even shut down the system if the temperature gets to an unsafe range.

8.8.2 Accelerometer

As its name suggests, the accelerometer sensor can be used to measure the acceleration exerted upon the sensor. Usually the acceleration is given in two or three axis-vector components that make up the sum/net acceleration. Accelerometers have quite a few uses. You can probably think of a few already—glass breakage detector, video game remote controls, or even electronic bubble levels for when you are trying to hang a picture frame on the wall. These accelerometers typically give us two types of data:

1. Static force applied on the sensor due to gravity → tilt/orientation detection
2. Force/acceleration exerted upon the sensor → movement/force detection

8.8.3 Microphone

Sound is another analog signal our LaunchPad can potentially read with one of its input analog pins. Depending on the frequency and shape of the sound wave, the microphone can convert this real-world wave into an analog voltage signal that represents different sounds and noises. Based on these analog readings, microcontrollers can usually extract amplitude or volume. However, depending on the capabilities of the microcontroller, even further information can be extracted from an audio analog signal.

Let's jump to the next project, where we will tinker with one of these analog sensors that just so happens to be available on our LaunchPad.

8.9 PROJECT 11: IS IT HOT IN HERE?

Let's play with an analog temperature sensor. Luckily for us, there's no need for any external components for this laboratory, so we will get to have some alone-time with our LaunchPad! The MSP430 microcontroller that is on our LaunchPad comes with an internal temperature sensor that is fed directly into an internal ADC channel. This means that we can use the `analogRead()` function to directly read the temperature that is being sensed by the MSP430 chip itself. Because the analog temperature sensor is integrated in the LaunchPad itself, our ingredient list is short.

Ingredients:

1. LaunchPad

   Let's open up a new Energia sketch and type the following code.

```
void setup() {
    Serial.begin(9600); // enable diagnostic data to be sent to
                        // the computer
    analogReference(INTERNAL1V5); // using the non-default
                                  // voltage reference
}

void loop()
{
    int analogValue;
    analogValue = analogRead(TEMPSENSOR); // read internal temp sensor
    Serial.println(analogValue,DEC); // send analog reading to computer
    delay(1000); // use this to delay new readings to once per second
}
```

OK, for the most part, pretty straightforward right? There's only one new function that we introduce in this example, which is the `analogReference()` function. If you recall from earlier in this chapter, we mentioned that the ADC has a few options to use as the voltage reference.

The MSP430 device on our LaunchPad has the following options:

- VCC (default)
- Internal high-precision 1.5 V reference → `analogReference(INTERNAL1V5);`
- Internal high-precision 2.5 V reference → `analogReference(INTERNAL2V5);`

The function `analogReference()` allows us to select which voltage reference to use that best fits our needs. What makes a voltage reference the right choice really depends on the application and several factors must be considered case by case to make such a decision.

Things you might look at when deciding on which voltage reference to use include:

• What is the expected voltage range that will be seen at the analog input pin and what kind of accuracy is needed for your application?
• Do you want absolute or relative measurement?

However, for most of our practical purposes, the rule of thumb is that if you are measuring an analog signal that is sourced by the same voltage source as the LaunchPad, then use the default VCC as your reference. The supply voltage VCC usually does not have the high-precision requirement as a dedicated voltage reference, and it can in fact fluctuate during runtime. Even though VCC may fluctuate during a sketch's operation, if both the LaunchPad and the source of the analog signal are provided voltage from the same VCC source, they should be relative to each other and using VCC as the reference will get you an accurate relative measurement. A good example for this would be the potentiometer, whose reading depends and scales with the supply voltage VCC. On the other hand, for analog components with output signals that are based on an absolute scale and do not fluctuate with changes of VCC, such as our temperature sensor, it is more accurate to use an absolute source such as our built-in 1.5 V or 2.5 V reference, which is more accurate than using VCC.

So, in summary, when the analog signal that we are reading scales appropriately with VCC, we're OK to use the default VCC voltage reference for our analog readings. Alternatively, if the analog signal we are reading is based on an absolute scale that doesn't scale with VCC, we are better off using one of the high-precision voltage reference options inside of our LaunchPad (1.5 or 2.5 V). In this example, we're using the 1.5 V reference voltage by passing in the parameter *INTERNAL1V5*.

Well, enough about our reference, let's move on to the next new concept introduced in this sketch, the "TEMPSENSOR" parameter. TEMPSENSOR allows the `analogRead()` function to select the appropriate analog input channel that is connected to the internal temperature sensor found in the MSP430 device on our LaunchPad. This is a predefined variable that is already inside of Energia that we can use as needed.

The `delay()` function slows down the rate at which the temperature sensor is read and the LaunchPad reports the data back to the PC. After all, temperature does not change that often or drastically, so we only need to see the updates once in awhile.

Let's download the sketch to the LaunchPad. Once it is up and running, open up the Serial Monitor to check the incoming messages.

You should see a stream of values like this coming in.

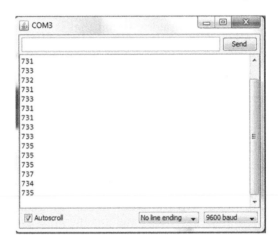

In order to see the changes in reading, you can either introduce a source of hot or cold to the surface of the MSP430 chip. A can of cold spray or an upside-down can of air dust spray can instantly cool down the surface temperature of the microcontroller. Alternatively, breathing onto or rubbing the surface of the LaunchPad can add some heat to the system. Try to either cool or heat the system and notice the change in returned values accordingly—we get higher values when hotter and lower values when colder.

Well, this correlation is neat and all but that does not tell you what the temperature is. We need more information to convert the raw voltage reading into a meaningful temperature reading.

The formula we can use for our temperature sensor is:

temperature in Fahrenheit = (analog reading − 630) × 761/1024;

This formula helps us convert our analog reading to a temperature value in Fahrenheit. This formula was derived based on the specification of the internal temperature sensor described in the MSP430G2553 device datasheet.

We use the code below to incorporate the algorithm above to convert our raw analog data into Fahrenheit. Let's plug this into our sketch.

```
void setup() {
        Serial.begin(9600);    // enable diagnostic data to be sent to the
                               // computer
        analogReference(INTERNAL1V5);   // using the non-default voltage
                                        // reference
}

void loop()
{
        long sensorValue = 0;
        long fahrenheitValue = 0;
        sensorValue  = analogRead(TEMPSENSOR)-630;
        fahrenheitValue = (sensorValue * 761) / 1024;
        Serial.print(fahrenheitValue, DEC);
        Serial.write('°'); // Serial.write() can be used tosend a single
                           // character/byte
        Serial.println("F");
        delay(1000); // use this to delay new readings to once per second
}
```

●●●──

Tip

The degree symbol (°) is not available on your typical keyboard. Two quick ways to get this key:

1. If you have internet access, simply search for "degree symbol" to find the degree symbol and copy it over to your sketch.
2. If your keyboard has a dedicated numeric keypad, hold down ALT and type in 0176, and ° should appear. What is this magic???? This is because ° is the 176th character in the ASCII table. What's ASCII, you might ask? Hold off your question for now, we don't want to reveal all the secrets on ASCII just yet. For the ASCII goodness, you'll have to wait until a future chapter, *The Languages of LaunchPad*. But for now, let's get back to our project.

Let's download this updated version of our sketch into our LaunchPad then head on over to our Serial Monitor.

You should be able to see the stream of more accurate temperature reading coming in similar to the figure below. Also, we can see we're using a clever combination of `Serial`.print(), `Serial`.write() and `Serial`.print() functions to make our data look nice and neat inside of our Serial Monitor!

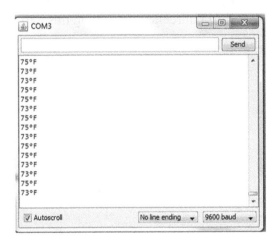

Congratulations on your first LaunchPad-powered electronic thermometer! You are now one step closer to solving global warming!

8.10 PROJECT 12: FUN WITH FORCES

In this project, we'll move on to yet another exciting analog signal. We introduced this sensor earlier on in the chapter, but now, let's get our hands dirty playing with the accelerometer!

Ingredients:

1. LaunchPad.
2. Educational BoosterPack
   OR
   Breadboard, ADXL335 breakout board

Today we will be playing with the three-channel (or three-axis) ADXL335 accelerometer. The accelerometer senses the total force or acceleration exerted on the chip and splits it into three vector components along the X-, Y-, and Z-axes. The three channels are connected to the three analog channels on the LaunchPad, which we can read separately. If you have your Educational BoosterPack handy, simply plug it into the LaunchPad. However, be sure to remove the jumpers that connect the LaunchPad's green and red LEDs (P1.0 and P1.6) so we don't interfere with any signals that connect to the BoosterPack.

Take a close look at the Educational BoosterPack. You will notice that two channels of the accelerometer share analog input pins with other analog components on the BoosterPack. Two jumpers can be used to ensure that our accelerometer signals are connected to the MSP430. Make sure to place the jumpers to the correct positions to ensure all X, Y, and Z signals get to their final destination as shown in the following image. The X channel is always connected to the LaunchPad; however, the Y and Z channels need to be routed with the two jumpers.

Connect Jumper J5 in the ACC_Y position

Connect Jumper J6 in the ACC_Z position

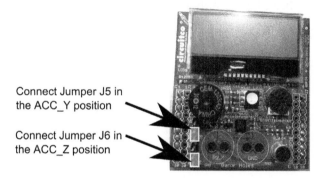

If you are not using the Educational BoosterPack, we can use a breakout board version of the ADXL335, which we can connect to our LaunchPad. Make sure the connections are wired up per the diagram below.

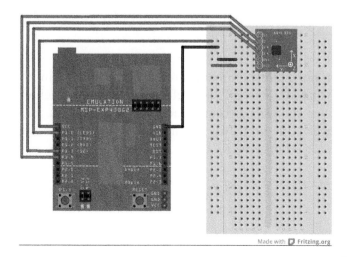

Made with Fritzing.org

As shown in the diagram, the accelerometer generates three analog signals—one for each axis—*X*, *Y*, and *Z*.

- The *X*-axis is connected to A0.
- The *Y*-axis is connected to A3.
- The *Z*-axis is connected to A4.

Now that we are done with the hardware, let's move on to the software.

Luckily for us, because the accelerometer is such a fun and popular sensor to interface it, there are already libraries and code examples written for the LaunchPad to interface with the device. Rather than reinvent the wheel, let's simply pull in one of these examples.

Go to *File > Examples > 6.Sensors > ADXL3xx.*

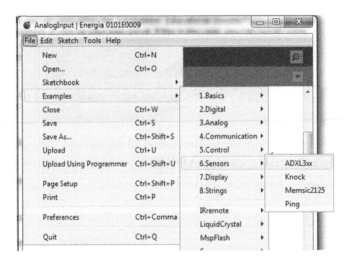

Spend a few minutes to read through the code. Again, there is not much new that you haven't seen before. Individually, each of the three accelerometer channels is being read using our `analogRead()` function. This is much in the same way we did for the other analog signals that we read in earlier projects.

There are only a few modifications we need to make to this example sketch to get it working with our particular accelerometer setup.

First off, let's fix the X, Y, and Z channel pin connections. From our hardware diagram, we see that X channel is connected to channel A0, Y to A3, and Z to A4. Let's edit the following lines of code:

```
const int xpin = A0; // x-axis of the accelerometer
const int ypin = A3; // y-axis
const int zpin = A4; // z-axis (only on 3-axis models)
```

Second, from the diagram, or schematic, we see that the accelerometer in our hardware is constantly powered by the VCC line and also directly connected to GND. Therefore, we don't need the power and ground pin functionality. We can safely remove the following lines of code at the top of the example:

```
const int groundpin = 18; //analog input pin 4 - ground
const int powerpin = 19; //analog input pin 5 - voltage
```

The reason this code example includes these lines of code is so that we can turn on the accelerometer as-needed by changing the states of these two I/O pins, which serve as the accelerometer's VCC and GND source. However, as our accelerometer in this case is "hard-wired" to VCC and GND, we don't need them. So let's go ahead and remove the associated lines of code in the **setup**() function as well.

```
pinMode(groundpin,OUTPUT); //remove this
pinMode(powerpin,OUTPUT); //remove this
digitalWrite(groundpin,LOW); //remove this
digitalWrite(powerpin,HIGH); //remove this
```

At this point, our complete code example should look like this:

```
// these constants describe the pins. They won't change:
const int xpin = A0;   // x-axis of the accelerometer
const int ypin = A3;   // y-axis
const int zpin = A4;   // z-axis (only on 3-axis models)

void setup()
{
        // initialize the serial communications:
        Serial.begin(9600);
}

void loop()
{
        // print the sensor values:
        Serial.print("X = ");
        Serial.print(analogRead(xpin));
        // print a tab between values:
        Serial.print("\t");
        Serial.print("Y = ");
        Serial.print(analogRead(ypin));
        // print a tab between values:
        Serial.print("\t");
        Serial.print("Z = ");
        Serial.print(analogRead(zpin));
        Serial.println(); // print out linefeed
        // delay before next reading:
        delay(100);
}
```

Now both our hardware and software are all set! You know the routine, click on the Verify and Download button in Energia! Once the sketch is running on the LaunchPad, fire up the Serial Monitor to check the incoming messages.

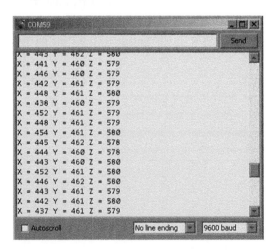

Now that we are streaming in accelerometer data, there are a few things that we should take note:

- Keep the LaunchPad very still, ensure no external force is exerted on the LaunchPad. When there's no force other than gravity exerted upon the device, the vector sum of the X-, Y-, and Z-axes should add up to be exactly equal to the gravity force, or $1g$. However, as these are vectors (magnitude and direction), we can't simply sum up each channel. We have to add them using vector math.
- Slowly tilt the LaunchPad from one direction to another, note the change in distribution between the channels. While moving or tilting the board, you are applying force to the board. But once it stops moving, the numbers should settle again, showing only the vector contributions of the $1g$ gravity force. If we are careful and perfectly still, as we tilt the LaunchPad, the three vectors should always add up to the $1g$ gravity force.
- Rapidly move the LaunchPad in different direction. Note the increase in values in all channels. By applying external force onto the device, you can increase the overall total acceleration or force applied to the device. In this case, when we are moving the LaunchPad, we are introducing additional forces along with the force of gravity. When external forces beyond gravity are introduced, the vector sum of the X, Y, and Z channels are then greater than $1g$ ($2g$, $3g$, etc.).

Spend a few minutes to play around with the sketch, hopefully you can come up with some ideas on how to leverage the measured acceleration to output or control something cool. As always, make sure to write your ideas down, whether they're possible at the moment or not. Who knows if later on you can come back and enable these ideas with the new things you learn throughout this book.

8.11 ANALOG OUTPUT

So far we have been learning about analog input. We now have the tools to convert an analog signal into digital 1s and 0s. We have also been playing with a few different types of analog inputs. Let's now move on to the other side of the fence, the analog output.

As its name suggests, an analog output generates a voltage level at the microcontroller pin. In a way, it is the complete opposite process when compared to analog input, as this time, a digital value is now being converted back into an analog voltage signal. Even though we're going in the opposite direction as the analog read process, most of the concepts and constraints are still applicable, such as the concepts of resolution, voltage step, and voltage range.

Resolution: While we are generating an analog output from the microcontroller, we already know that the microcontroller is only capable of creating a finite number of analog outputs due to its digital brain. However, the higher the resolution, the closer we can get to creating an idealistic analog output.

Voltage Range: Similar to the analog input, the analog output is usually capable of generating a voltage signal within the range of 0 and the reference voltage, which is VCC by default.

Voltage Step: Again, this can be derived by Voltage Range/Resolution, yielding the minimum voltage width or step that the analog output can generate.

Without further ado, let's try out a quick sketch to output an analog signal.

8.12 PROJECT 13: ANALOGWRITE() WITH ONE LED

For this project, we will only need the LaunchPad and one of its onboard LEDs. That's right, we're back to blinking LEDs! However, this will be a bit different since we'll be adding in a hint of analog. In this project, we're going to output a variable voltage on pin 14, which is what connected to the green LED on our LaunchPad. By changing the output voltage on pin 14, we should be able to directly change the brightness of the green LED. Be sure the jumper on J5 is attached to ensure an electrical connection between the green LED and pin 14.

Ingredients:

1. LaunchPad

Open up a new Energia sketch and add the following code.

```
void setup() {
    pinMode(14, OUTPUT);
}
void loop() {
    for (int i=0;i<256;i=i+5){
    analogWrite(14, i); // output analog signal on green LED (pin 14)
    delay(30);
    }
}
```

In this project, we're introducing yet another function called analogWrite(). Before we download this new code to our LaunchPad, let's take a moment to learn about this new function. The function actually looks a lot like our digitalWrite() function. Much like digitalWrite(), analogWrite() accepts two parameters. The first parameter tells the function which output pin we should use to generate our analog signal. The second parameter tells the function what level the analog voltage output should be.

In the code example, we are outputting the analog signal through pin 14, which is connected to the green LED. We are also using a for-loop that increments from 0 to 255, five steps at a time.

In this example, we're actually using the counter "i" from our for-loop in a clever way. We are passing it into the analogWrite() function. Why?

Well, by passing in the value "i" into our analogWrite() function, we are outputting a new analog value that will grow from 0 to 255, until the for-loop completes. Once the counter "i" in our for-loop is no longer <256, our sketch exits the for-loop and we return back to our main loop(), which will then repeat itself. When the main loop() function repeats itself, the for-loop will reinitialize and run through its course again, outputting an analog voltage level that grows from 0 to 255. And as usual, this loop() function will continue to repeat forever causing our green LED to start from the OFF state and gradually get brighter and brighter, until finally it becomes fully ON and loops back around and restarts at the OFF state again.

The reason our for-loop is bounded between 0 and 255 is because this is the range that our analogWrite() function is capable of outputting. This means that the analogWrite() function available for our

LaunchPad has a resolution of 8 bits since the largest number we can represent with 8 bits is $11111111_2 = 255_{10}$.

The `analogWrite()` function can output a voltage range between 0 V and VCC (~ 3.6 V). The digital number that we pass into the `analogWrite()` function is then mapped from 0–255 to a voltage output of 0 V to ~ 3.6 V (VCC).

Since we're using a for-loop to increment the output from 0 to 255, we should see the voltage output on pin 14 start at 0 V, then increase to VCC (~ 3.6 V). As pin 14 is connected to our green LED, the LED should start with being OFF and gradually increase in brightness as its supply voltage rises.

So let's put it to the test. Click Verify and Download in Energia to see what happens to the green LED on the LaunchPad.

At this point you should be able to observe and confirm that the LED output gradually increases from OFF to ON and repeats the cycle again. Nice work!

Once we see our LED going through this modified blinking example, let's head back to our Energia sketch to tidy up our code a bit.

8.13 PROJECT 14: HOUSE CLEANING AND CODING BEST PRACTICES

You might recall from previous sketches, we usually define the pins, whether analog or digital, with the help of variables. In a way, variables can allow us to give each pin a nickname. So instead of directly addressing a pin by its number, we can address it by its nickname "GREEN_LED." This is useful because we can use these nicknames/variables throughout the rest of the sketch. This makes the code more understandable and organized.

Imagine if you want to change the previous project so that a different LED glows instead of the green one (pin 14). If the pin number is defined and given a nickname/variable at the beginning of the program, you'll simply have to redefine the nickname just one time with the new pin number assignment. On the other hand, if you hard-coded the pin number throughout the sketch, like what we hastily did in our previous project, you'll end up spending lots of time going back changing each and every instance. Well, luckily for us, we only hard-coded the pin number 14 in two locations.

So let's do our due diligence and go back to tidy up the sketch, and hopefully learn the lesson that we should not let the immediate excitement to get something up and running overcome the need for organized code.

1. First, let's add the following line of code before the `setup()` function to give our analog pin a nickname using the variable "analogPin":

```
const int analogPin = 14; // assign Pin #14 to analogPin
```

Note that for `analogInput()`, we used analog channels (A0, A1, A2, ...). But for the `analogOutput()` function, we are simply using the pin number designation.

2. Replace the number 14 in the `pinMode(14, OUTPUT)` and `analogWrite(14, i)` function call with our new nickname for pin 14:

```
analogWrite(analogPin, i);
```

3. Redownload and confirm the LED behaves the same as before.

Tip

Try to avoid hard-coding numbers of values into your code. Instead, for numbers or values that are used multiple times in the sketch, define them at the beginning of the program. This way, when you have to update or change those definitions, you'll only have to change it once.

8.14 PROJECT 15: HEART BEAT

In our current sketch, the LED gradually turns on and gets brighter but then abruptly turns OFF and restarts the cycle. It would be nice to smooth this transition out by first gradually increasing the LED's brightness until it's fully ON and then gradually decreasing the brightness all the way down to OFF. That way, for each loop, we complete a full cycle of OFF → ON → OFF. Do you think you can tackle this one by yourself?

Give it a shot and once you've modified your code, compare it against the solution below.

Solution: Add a reverse for loop after the first for-loop, so that together they'll look like this:

```
const int analogPin = 14;          // assign Pin #14 to analogPin

void setup() {
      pinMode(analogPin, OUTPUT);
}

void loop() {
      int i;
      for (i=0;i<255;i=i+5){
            analogWrite(analogPin, i);
            delay(30);
      }
      for (i=255;i>0;i=i-5){
            analogWrite(analogPin, i); // notice that our
                                       // nickname came in handy!
            delay(30);
      }
}
```

One last enhancement. Try to simulate a heartbeat with the LED fading in and out. A human's resting heart rate is between 50 and 80 beats per minute If you feel like our simulated heart beats too fast or too slow, you might want to adjust the delay between each analogWrite() step.

How about simulating other animals' heartbeat? A blue whale's heart beats 6 times per minute, while a rabbit's heart beats 200 times per minute.

8.15 PROJECT 16: MIXING COLORS WITH AN RGB LED

Let's continue to improve upon our basic blink LED example with the help of our new analogWrite() function. In Project 15, we used analogWrite() to change the voltage output of a single LED, which we could perceive as changes in the green LED's brightness. In this laboratory, we're going to be using something called an RGB LED. An RGB LED is unique because it's essentially three LEDs in one! Inside of a single RGB LED is a **R**ed, **G**reen, and **B**lue LED. With the help of our analogWrite() function, we are going to change the brightness of each of these channels, which will cause the three colors to mix together creating a wide array of new colors!

Ingredients:

1. LaunchPad
2. The Educational BoosterPack
   OR
   Breadboard, RGB LED, 2×100 Ohm resistor and 1×150 Ohm resistor

The Educational BoosterPack comes with an **RGB LED**, so if you have the BoosterPack, you're good to go. As usual, before plugging the BoosterPack onto the LaunchPad, make sure to remove the LaunchPad's onboard **LED** jumpers and populate the **TX** and **RX** jumper for our Serial communication channel to enable debugging.

If you are building your own breadboard circuit, you will need to get few more components as listed above. Wire up the LEDs on the breadboard to the LaunchPad as shown in the diagram below.

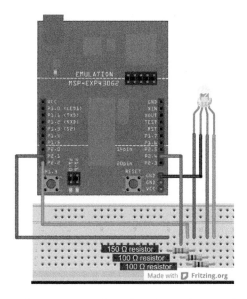

The key connections to keep in mind for this exercise are:

- Red LED—Pin 9
- Green LED—Pin 10
- Blue LED—Pin 12

The framework for this sketch will be pretty much the same as the previous sketch, except repeated 3 times for each of the three color channels of the LED.

1. So if you'd like, you can simply copy and paste the code from Project 15 just to start off. Once you have the code copied over, we'll need to go through to update and replicate some lines of code for the three channels.
2. The first step would be the pin definition on the very top.
   We need to make new nicknames for the three pin assignments:
   Remove:

```
constintanalogPin = 2; //assignPin#2toanalogPin
```

Add:

```
const int redLED = 9; // assign red LED to Pin #9
const int greenLED = 10; // assign green LED to Pin #10
const int blueLED = 12; // assign blueLED to Pin #12
```

3. In the `loop()` function, replace the content of both for-loops with:

```
analogWrite(redLED, i);
analogWrite(greenLED, i); .
analogWrite(blueLED, i);
```

The final product should look like this.

```
const int redLED= 9;        // assign red LED to Pin #9
const int greenLED= 10;     // assign green LED to Pin #10
const int blueLED= 12;      // assign blueLED to Pin #12

void setup() {
        pinMode(redLED, OUTPUT);
        pinMode(greenLED, OUTPUT);
        pinMode(blueLED, OUTPUT);
}

void loop() {
        int i;
        for (i=0;i<255;i=i+5){
                analogWrite(redLED, i);
                analogWrite(greenLED, i);
                analogWrite(blueLED, i);
                delay(30);
        }
        for (i=255;i>0;i=i-5){
                analogWrite(redLED, i);
                analogWrite(greenLED, i);
                analogWrite(blueLED, i);
                delay(30);
        }
}
```

Note

Before we fire off the Verify and Download button in Energia, be aware that the RGB LED can be very bright, so take the appropriate precautions! Placing a blank white piece of paper on top of the LED does a decent job diffusing the light and protecting our eyes!

Go ahead and verify, download, and fire up the sketch on your LaunchPad and see the synchronous RGB LED heartbeat.

"What do you mean synchronous?" you might ask. The three R, G, and B LED channels are being updated at the same time AND being changed in the same way; they are assigned the same value "i" which causes each channels' brightness to go up and down in unison.

OK that's cool and all, but we promised a lot of colors! Instead, we're just seeing the red, green, and blue light combine together with equal intensity, which creates a pulsating white light.

Well, let's mix it up then! We need to make sure that the values for the three channels are not related and they will be changed in different ways from each other. This is also a good opportunity to introduce another cool function that comes with Energia, `random()`.

As its name suggests, the `random(maxValue)` function returns a random value between 0 and maxValue. You can also call `random(minValue, maxValue)` to get a random value between the (minValue, maxValue) range.

As our `analogWrite()` output resolution is 256, we'll use the range (0,255) as the input parameters for the `random()` function.

Before we can use the `random()` function, we first need to generate a random seed at the beginning of the sketch. This ensures that the values are random and change every time the sketch runs.

In the `setup()` function, add the following line. To generate a random seed, we will read in the noise coming in through analog channel A0. The noise that we read will help us generate a random seed:

```
randomSeed(analogRead(A0));
```

Once we have the random seed generated, we can then use the `random()` function in the rest of our sketch. Let's go back to Energia and replace everything inside of the `loop()` function with the following code.

```
void loop() {
        analogWrite(redLED, random(255));
        analogWrite(greenLED, random(255));
        analogWrite(blueLED, random(255));
        delay(500);
}
```

Give this sketch a try. You should now notice that even though all three RGB channels are being changed at the same time (still time-synchronous), their values are being changed differently, resulting in different combinations of colors.

Let's take this a little bit further. How about this time, let's make the RGB channels change their values at different times, independently from each other.

Replace the void `loop()` function with the following.

```
void loop() {
        analogWrite(redLED, random(255) );
        delay(random(50,500));
        analogWrite(greenLED, random(255));
        delay(random(50,500));
        analogWrite(blueLED, random(255));
        delay(random(50,500));
}
```

In this new version of the code, the channels are updated independently, and their updates are separated by a delay of random length. Notice that we are reusing the `random()` function to randomize the delay duration between each LED channel update. Pretty nifty, huh?

Give this new sketch a go. What do you think of your LaunchPad-powered disco ball?

If it is not up to your liking, feel free to further tweak the parameters such as the maxValue for the random functions or the delay durations.

8.16 ANALOG-TO-ANALOG SIGNAL CHAIN

Now that we understand both ends of the analog signal: input and output, we have the capability to combine them together to create a complete analog-to-analog signal chain. The beauty of this signal chain is that our LaunchPad with its digital brain is caught in the middle of it! By being the middle-man in an analog-to-analog signal chain, the LaunchPad can simply pass through, adjust, or modify the original analog signal to ouput one that might look completely different.

Let's close this chapter with a couple more sketches that tie together the different analog concepts that we learned in this chapter.

8.17 PROJECT 17: ANALOG-TO-ANALOG SIGNAL CHAIN: POTENTIOMETER → WHITE LED

The goal of this project is to use the potentiometer analog input to control the analog output at the LED pin.

Ingredients:

1. LaunchPad
2. Educational BoosterPack
   Or
   Breadboard, 10,000 Ω potentiometer

If you are using the Educational BoosterPack, we will use the white LED for this exercise. As usual, make sure to unpopulate the LaunchPad's onboard LED jumpers before plugging the BoosterPack onto the LaunchPad. Also, place the RXD and TXD jumpers appropriately if you want to enable communication of diagnostic data to the computer for simple debugging.

If you are wiring up your own breadboard circuit, make sure to follow the diagram on the next page.

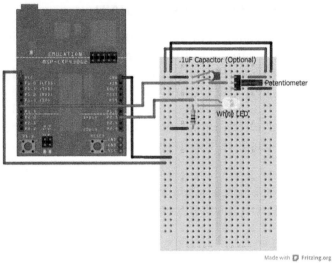

Made with **D** Fritzing.org

If you are using the Educational BoosterPack, there is an important connection that we should point out here. The potentiometer is connected to pin 5, which is our analog channel 3 (A3). The Y channel of our accelerometer on the BoosterPack is also connected to channel A3. To ensure that the potentiometer is connected jumper J5 on the BoosterPack needs to be populated in the A3_POT position.

Also, for those using the Educational BoosterPack, the white LED is connected to P2_2. For those using the LaunchPad + breadboard approach, you have a choice to use P2_2 for an external LED, or you can simple use the green LED that's already available on the LaunchPad, which is connected to pin 14.

Awesome, hardware is complete! Let's switch gears and start writing some code.

Open up a new Energia sketch and type in the following code.

```
const int potChannel = A3;       // assign potentiometer to channel A3
// ****Comment out one of the following lines *****
// ****Depending on which circuit you use with your LaunchPad *****
const int ledPin= P2_2; // for Educational BoosterPack user,
                        // use this line (Green LED)
//const int ledPin =14; // if using LaunchPad's on-board green
                        // LED, use this line

void setup() {
        pinMode(ledPin,OUTPUT);
}

void loop() {
        int analogValue;
        analogValue = analogRead(potChannel);
        analogWrite(ledPin, analogValue);
}
```

OK nothing fancy here, we're simply passing the value measured from the potentiometer along to the LED. Let's see what happens on the LaunchPad shall we? You know the drill: click Verify and Download to see the LaunchPad in action.

By turning the potentiometer, you should be able to adjust the brightness of the LED. If you dial the potentiometer down to 0, the LED should be completely off.

But as you dial the potentiometer up, you will soon notice that at some point, even though you can still turn the potentiometer, the LED cannot get any brighter. What's going on here?

The reason behind this is the difference between the resolution of the analogRead() and the resolution of the analogWrite() functions. If you recall, analogRead() returns a value between 0 and 1023, while analogWrite() can only output a value between 0 and 255. 255 represents the maximum output voltage possible since our analogWrite() has a resolution of only 8 bits, while analogRead() has a resolution of 10 bits.

So how do we fix this? To properly represent the output and the input on the same relative scale, we will need to scale the larger resolution down to the smaller resolution. In our case, we will need to map the `analogRead()` return value down from (0,1023) down to (0,255). Conveniently enough, this is simply accomplished by dividing the value by 4. Which leads us to updating the following line of code:

```
analogWrite(ledPin, analogValue / 4);
```

Go ahead verify and download the updated sketch to see if the full range of the potentiometer correlates to the full range of brightness on the LED. Aaaahhh, that's better!

8.18 PROJECT 17: ANALOG-TO-ANALOG SIGNAL CHAIN: ACCELEROMETER → RGB

This project is a slight modification of Project 16. In this project, we are going to use the accelerometer readings from the X, Y, and Z channels and map that to our RGB LED. Each accelerometer channel should control the brightness of each LED. This way, by tilting our LaunchPad in different directions, we should be able to generate all sorts of colors!

Ingredients:

1. LaunchPad
2. Educational BoosterPack
   OR
   Breadboard, ADXL335 breakout board, RGB LED, $2 \times 100 \, \Omega$ and $1 \times 150 \, \Omega$ resistor

As we have both an ADXL335 accelerometer and an RGB LED on our Educational BoosterPack, setup will be very easy for those that have that handy. Alternatively, we can also assemble this on a breadboard.

For those with the Educational BoosterPack, simply plug the BoosterPack onto the LaunchPad. Be sure to remove the LED jumpers and place the TXD and RXD jumpers properly to enable debug on the LaunchPad. Once the BoosterPack is plugged in, make sure that the two jumpers on the BoosterPack are in place to ensure that the Y and Z channels of the accelerometer are connected to the LaunchPad.

For those that will put together the circuit on a breadboard, follow the diagram below.

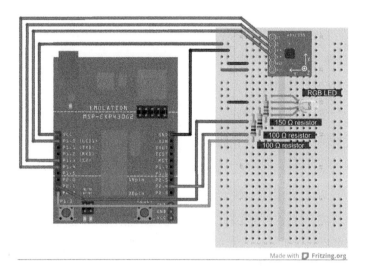

Made with **F** Fritzing.org

As you may notice, this is a combination of two projects that we did earlier in this chapter. Once you have your hardware setup, let's turn our attention to Energia and write some software.

●●●——————————————————————————————

Tip

This is a long piece of code, so if you don't feel like retyping 3 pages worth of code, you can go to the http://booksite.elsevier.com/9780124115880 and download the sketch in its entirety. We have placed the code in the book here for your reading convenience.

```
const int xpin = A0;   // x-axis of the accelerometer
const int ypin = A3;   // y-axis
const int zpin = A4;   // z-axis (only on 3-axis models)

const int redLED = 9;   // assign red LED to Pin #9
const int greenLED = 10;// assign green LED to Pin #10
const int blueLED = 12; // assign blueLED to Pin #12

void setup() {
    Serial.begin(9600);        // initialize for sending diagnostic
                               // info to computer
}

void loop() {
    int analogValue;

// CHECK X AXIS (red)
    analogValue = analogRead(xpin); // read X axis
    Serial.print("RED (X): ");      // print to serial monitor
    Serial.print(analogValue);
    Serial.print("\t");
    if(analogValue >490){      // check if tilting on x axis
                               // in positive direction
    analogValue = map(analogValue, 490,600, 0, 150);  // map to RGB range
    analogWrite(redLED, analogValue); // output color
    }
    else if(analogValue<430){ // check if tilting on x axis
                               // in negative direction
    analogValue = map(analogValue, 340, 450, 150, 0);  // map to RGB range
    analogWrite(redLED, analogValue); // output color
        }
    else{
    analogWrite(redLED, 0);// analogValue is b/w 450 & 490,
                           // X axis is ~flat
    }

 // CHECK Y AXIS (green)
    analogValue = analogRead(ypin); // read Y axis
    Serial.print("GREEN (Y): ");    // print to serial monitor
    Serial.print(analogValue);
    Serial.print("\t");
    if(analogValue >490){      // check if tilting on Y axis
                               // in positive direction
    analogValue = map(analogValue, 490,600,0,150);  // map to RGB
                                                    // range
    analogWrite(greenLED, analogValue);        // output color
    }
    else if(analogValue<450){       // check if tilting on Y axis
                                    // in negative direction
    analogValue = map(analogValue, 340,450, 150, 0);  // map to RGB
                                                      // range
    analogWrite(greenLED, analogValue); // output color
        }
```

```
        else{
        analogWrite(greenLED, 0);// analogValue is b/w 450 & 490,
                            // Y axis is ~flat
        }
// CHECK Z AXIS (blue)
        analogValue = analogRead(zpin);   // read Z axis
        Serial.print("BLUE (Z): ");       // print to serial monitor
        Serial.print(analogValue);
        Serial.print("\t");
        if(analogValue >490){             // check if tilting on Z axis
                                          // in positive direction
        analogValue = map(analogValue, 490, 600, 0, 150);  // map to RGB range
        analogWrite(blueLED, analogValue); // output color
        }
        else if(analogValue<450){ // check if tilting on Z axis in negative direction
        analogValue = map(analogValue, 340,450, 150, 0);  // map to RGB range
        analogWrite(blueLED, analogValue);      // output color
        }
        else{
        analogWrite(blueLED, 0);  // analogValue is b/w 450 & 490,
                            // Z axis is ~flat
        }

}
```

In this Energia sketch, we are checking each axis of the accelerometer one at a time. We use the `analogRead()` function to check each analog channel—one for each accelerometer axis. Then, we use a few if-else-statements to check which direction the accelerometer is being tilted toward. Once we know if we are tilting along a particular axis in the positive or negative direction, we use the `map()` function to scale the analog input signal into an appropriate analog output value, which we finally send to the `analogWrite()` function to set the specific color on one of the R, G, or B LEDs.

Note

Before we fire off the Verify and Download button in Energia, be aware that the RGB LED can be very bright, so take the appropriate precautions! Placing a blank white piece of paper on top of the LED does a decent job diffusing the light and protecting our eyes!

Go ahead and download this sketch into the LaunchPad by pressing the Verify and Download button in Energia. Once it starts running, tilt the accelerometer along the three axes. The image below shows the direction of each axis.

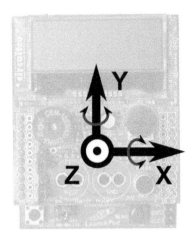

As we tilt the accelerometer sensor along each axis, we should see the individual R, G, and B channels change intensity.

- When the sensor is laying flat on the table, the X- and Y-axes are flat and no tilt is sensed. Thus, only the Z-axis has any force exerted on it (this is the natural force of gravity). As the Z-axis is represented by the blue channel, that's the only color we should see.
- When we start to tilt the board along the X-axis starting in a flat position, either in the positive or negative direction, we should see the color change smoothly from blue to purple to red.
- When we start to tilt the board along the Y-axis starting in a flat position, either in the positive or in the negative direction, we should see the color change smoothly from blue to teal to green.

And that's all there is to it! We created a simple analog-to-analog signal chain. Go ahead and play around with this sketch by modifying the map function or try switching up the way the input channels (accelerometer) pair up with the output channels (RGB LED).

8.19 ANALOG WRAP UP

In this chapter, we explored the world of analog. We learned that the LaunchPad can take in analog signals of infinite resolution from the external "real world" and convert it into the digital numbers that it can recognize and process. Our LaunchPad is also able to create analog signals of finite resolution to communicate back with the external world.

In the next chapter, we'll actually learn a little secret about the LaunchPad and its ability to generate "analog" signals!

Also, be sure to continue thinking about the dream application that we hoped to create at the very beginning of our journey. With this new understanding of analog signals, write down any analog inputs or outputs that you think you may need to accomplish your goals in the space below.

CHAPTER 9

1s and 0s Revisited: The Digital Stream

9.1 THE FLEXIBLE DIGITAL WORLD

In the previous chapter, we learned about how the LaunchPad can interact with the vast and colorful world of analog. When you compare analog's world of infinite colors to the black and white world of digital, one might come to the conclusion that analog is greatly superior. Well, let's put that conclusion away because we have a few clarifications to make.

It is true that 1s and 0s are quite limited in terms of the information they can carry individually. It is also quite unfortunate that the microcontroller's brain can only understand and interpret these 0s and 1s. But don't forget the fact that the LaunchPad brain can digest and process those 0s and 1s very quickly. Millions or even up to billions of computations can happen every second depending on the microcontroller. Our LaunchPad can do up to 16 million computations made up of these 0s and 1s every second.

Thanks to this speed, our LaunchPad can actually send out a stream of 1s and 0s in various patterns to create more information beyond just ON and OFF, HIGH and LOW, or black and white. The first method that we'll introduce is PWM or Pulse-Width Modulation.

9.2 PULSE-WIDTH MODULATION

Define

Pulse-Width Modulation

Pulse-width modulation is a digital technique to control a signal by repeatedly toggling a signal between a HIGH and a LOW state in a consistent pattern. We can portray new information by changing how long the signal is HIGH versus LOW. The PWM signal has two key parameters—frequency and duty cycle.

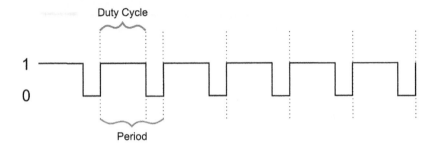

The period of the PWM signal (measured in seconds) indicates the amount of time it takes for the signal to complete one cycle. Refer to the image above. We can see that a period is a complete cycle of the signal, changing from LOW signal to a HIGH signal and back to a LOW signal. The signal pattern of each period is repeated over and over again.

The *frequency* of the PWM signal, measured in Hertz (Hz), indicates the number of complete periods that occur in 1 second. In our LaunchPad, the PWM signal typically has a frequency of 490 Hz. This means that our PWM signal completes 490 periods every second. The period and frequency of a signal are closely related, in that the faster the frequency, the smaller the period. The slower the frequency, the larger the period.

The relationship between period and frequency is realized with this equation:

$$\text{Period(s)} = \frac{1}{\text{Frequency(Hz)}}$$

For example, the period of a 1-kHz (1000 Hz) signal is 1/(1000 Hz) = 0.001 s or 1 ms.

The *Duty Cycle* of the PWM signal is the percentage of time that our PWM signal is in a HIGH state versus a LOW state. For example, a simple square wave has a duty cycle of 50%. However, we can change the duty cycle of our PWM signal, anywhere from 0% (for always OFF) to 100% (for always ON) and anywhere in between, depending on what type of information we want to portray.

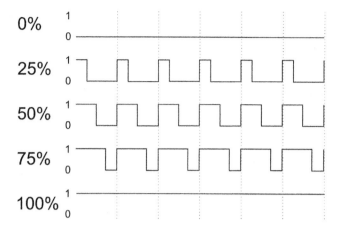

And that's it! PWM is a simple digital signal and is a great example of how we can start to extract more information out of the digital brain of our LaunchPad.

PWM is primarily useful because of the ability to choose the duty cycle from a fixed frequency digital signal. Once fixed on a specific frequency, 490 Hz in our example, the LaunchPad can convey different information to other "intelligent" devices by varying the duty cycle from 0% to 100%. The PWM signal receiver can decode the duty cycle into different available actions and behave accordingly.

Actually, PWM can also be used to control the power that is fed into certain electrical devices. By changing the duty cycle of a PWM signal, we can actually simulate an *average* voltage (or current) output. The longer the duty cycle is, relative to the period, means that, on average, the signal is more HIGH than LOW. This means on average more voltage (or current) is seen at the digital output pin. If we start at a duty cycle of 0%, we should see an effective average voltage of 0 V. Increasing duty cycle to 50%, we'll start to see an effective average voltage of ∼1.8 V (VCC/2). And lastly, at a duty cycle of 100%, we should see an effective voltage of ∼3.6 V (VCC).

As a matter of fact, this is exactly how we created the analog output signal in our previous chapter using our `analogWrite()` function.

9.3 WHAT? ANALOGWRITE() IS A LIE

Sadly so. The MSP430 device on the MSP430 LaunchPad does not actually have a true analog output module, which is often called a digital-to-analog converter (or DAC). Now, some other MSP430 devices do have the actual DAC modules, but the MSP430G2553 device on our LaunchPad does not.

If we take a close up look at our PWM signal, the ON and OFF components inside of a single period look very discrete and separated. However, the faster the frequency of our PWM signal, the harder it is to discern between the ON and OFF state and the signal appears to be averaged out. To explain this concept, think of an animated flipbook. If we flipped this animation one time per second, our animation would look jerky and very unconvincing. However, if we increase the frequency of our flipbook to change frames 60 times per second, suddenly the animation is smooth and realistic. Similarly, we can simulate an average voltage output at a pin by varying the duty cycle at a pin when frequency is high enough. Fortunately for us, the default 490 Hz frequency of our LaunchPad is quick enough to fool most external devices into thinking a stable voltage is present.

A component can perceive the PWM signal as an average voltage signal that is somewhat between VCC and 0 V. The value of this averaged voltage signal can be determined by the relative ON duration of the signal in each period or cycle. For example, if the duty cycle of the signal is 75%, the averaged out voltage level can appear to be ~75% of VCC.

$$\text{Average signal (V)} = \text{Duty cycle (\%)} \times \text{VCC (V)}$$

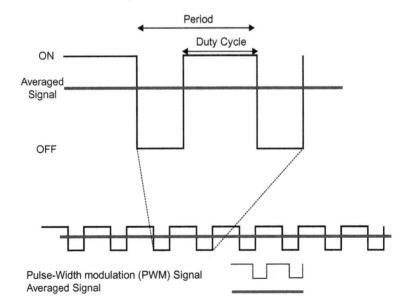

So just to reiterate, the `analogWrite()` function that we played with last chapter actually isn't analog at all! It's truly a digital function that simulates an analog output by varying the duty cycle of a PWM output signal.

The parameter that we passed into our `analogWrite()` function was simply telling the PWM signal what duty cycle it should toggle at. Because our `analogWrite()` function can accept an input parameter between (0–255), we can tell our PWM signal to generate up to 256 unique duty cycles. An input of 255 causes a duty cycle of 100%, while an input of 0 creates a duty cycle of 0%. This input maps linearly and we can get a good idea of what average voltage we can expect to see at the digital output pin for `analogWrite(analogWriteInput)` using the equation below:

$$\text{Average voltage} = \text{VCC} \times (\text{analogWriteInput}/255)$$

So we were using PWM signals without even knowing it! In Chapter 8, we used this digital trick to change the intensity of LEDs. We even used three PWM signals to change the intensity of three LEDs to create a wide array of colors.

As we have actually spent quite some time with our PWM signal between this chapter and the last, let's spend the rest of the chapter learning more about a special type of PWM signal—the square wave.

9.4 PROJECT 17: SQUARE WAVE AND A BUZZER

A square wave is a unique PWM signal which has a duty cycle of 50%—this means that the signal is ON just as much as it is OFF. In this project, we will use a square wave to make some noise using a piezo buzzer.

Ingredients:

1. LaunchPad
2. Educational BoosterPack
   OR
   Breadboard, piezo buzzer (2.048 kHz), 33 Ω resistor

Piezo buzzers are used for making bleeps, bloops, and tones. We can generate these types of sounds by providing a square wave to a piezo buzzer. The frequency of the square wave (Hz) determines the pitch of the noise the buzzer generates. The slower the frequency, the lower the pitch. The higher the frequency, the higher the pitch.

The buzzer that comes with the Educational BoosterPack is pretty small, but it does pack quite a punch. If you have the Educational BoosterPack, simply plug it into the LaunchPad. As always, be sure to have the LED jumpers removed (P1.0 and P1.6) and position the TXD and RXD jumpers to enable the LaunchPad to send diagnostic information to the computer for simple debugging.

If you are not using the Educational BoosterPack, you can wire up an external buzzer as shown in the following diagram. Connect one pin to ground and the other pin to a PWM-capable pin on the microcontroller (the piezo is bi-directional so either pin can be connected to GND or the MSP430). In this example, we are going to connect it to the XIN pin (pin 19). This is exactly the same way the buzzer on the Educational BoosterPack is connected to our LaunchPad.

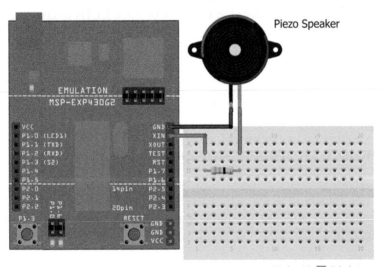

Made with **Fritzing.org**

●●●

Tip

Also, refer to the pin-mapping guide that is available in the Appendix. The pins that are "purple" have the ability to generate PWM signals using the `analogWrite()` function.

Once your hardware is all set, open up a new sketch inside of Energia and type in the following lines of code.

```
const int buzzerPin = 19;
void setup() {
}

void loop() {
  tone(buzzerPin, 698, 500); // 698Hz = Note F5. Hold note for 500ms
  delay(500);                // delay between tones
}
```

Once done typing, click Verify and Download. Once it's loaded up, the sketch should play the note F5 for half a second, pause for half a second, and repeat the tone again.

There is a new function that is introduced in this sketch. The `tone (buzzerPin, 698, 500)` function call instructs the LaunchPad to generate a 698-Hz tone for 500 ms at the buzzerPin. This `tone()` function is very similar to our `analogWrite()` function. The main difference, however, is that the parameter in `analogWrite()` changed the duty cycle while frequency stayed constant. With `tone()`, our parameter changes the frequency, while duty cycle stays constant at 50%. We also get one more parameter to play with, which tells the `tone()` function how many milliseconds it should hold the note for. Unlike most, our LaunchPad can hold a pitch quite well!

The time parameter is an optional one. If the time parameter isn't provided, then the specified pitch will be held indefinitely.

If a time parameter isn't provided, we need to use another function to stop the square wave. The `noTone(buzzerPin)` can be used to tell the LaunchPad to stop with all the noises that might be sourced by the buzzerPin.

9.5 PROJECT 18: DOE, A DEER, A FEMALE DEER (MUSIC WITH LAUNCHPAD)

What we created in Project 17 is a single-tone generation. Let's extend on this and create a multiple tone sketch. As we have found out before, our piezo buzzer is capable of generating different tones. To be exact, we can generate tones that are within the (human) audible frequency range (~100 Hz to a few kiloHertz). Furthermore, we also have the freedom to specify the duration of each tone. Combining the components of specific notes and duration, along with a tiny hint of musicality, we can make ourselves a short melody.

For this project, we will use an existing example sketch available in Energia. The sketch can be found in the following path:

File > Examples > Digital > toneMelody.

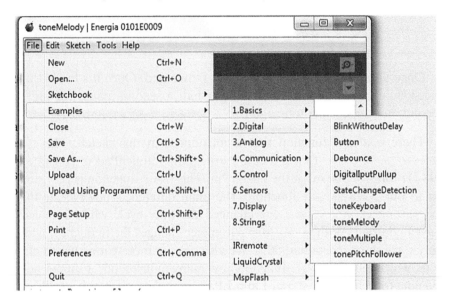

The first thing to note when you open this example is that two files are opened. We have our main code example, but we also have a second file that is hidden away in a different tab inside of Energia.

Our main code file is in the tab labelled *toneMelody* and the second file *"pitches.h"* is in the second tab. This second pitches.h file provides our project with additional information and capabilities that gets pulled into the sketch.

We can pull in the added information or functionality from pitches.h by including the following line of code in the main toneMelody sketch:

```
#include "pitches.h"
```

The #include keyword allows us to bring in additional functions and definitions that are defined in specified *.h file. In this case, the pitches.h file that we are including pulls in specific definitions and functions that serve a specific purpose. These types of files are usually called *libraries*. To make our music-making easier, we are importing this pitches.h library, which will give us unique functions and definitions for creating specific musical notes.

Packaging common functionalities into a library allows it to be reused by different sketches and applications. In our example, we are pulling in the information from "pitches.h." If you switch over to the pitches.h tab, you will find out that the file contains all the musical note frequency definitions. This allows us to easily pull in the correct frequencies for desired musical notes for our melody.

Let's switch back to the toneMelody tab and have a quick rundown of the rest of code. Here's the sketch in its entirety.

```
#include "pitches.h"
// notes in the melody:
int melody[] = {
    NOTE_C4, NOTE_G3, NOTE_G3, NOTE_A3,
    NOTE_G3,0, NOTE_B3,NOTE_C4};
// note durations: 4 = quarter note, 8 = eighth note, etc.:
int noteDurations[] = {4, 8, 8, 4, 4, 4, 4, 4 };
void setup() {
    // iterate over the notes of the melody:
    for (int thisNote = 0; thisNote < 8; thisNote++) {
        // iterate through melody
        // to calculate the note duration, take one second (1000 ms)
        // divided by the note type.
        // e.g. quarter note = 1000 / 4,
        // eighth note = 1000/8, etc.
        int noteDuration = 1000/noteDurations[thisNote];
        tone (8, melody[thisNote],noteDuration);
        // play specified note
        // to distinguish the notes,
        // set a minimum time between them.
        // the note's duration + 30% seems to work well:
        int pauseBetweenNotes = noteDuration * 1.30;
        delay(pauseBetweenNotes);
        // stop the tone playing:
        noTone(8);
    }
}

void loop() {
    // no need to repeat the melody.
}
```

Note the two array definitions of melody[ ] and noteDurations[ ]. These two arrays of equal size go hand in hand to define the note frequency and the associated duration for each note of the melody. Wait, what is an array?

Define

Array

An *array* is a collection of data or variables. An array is useful because the dataset that is contained inside of the array is indexed. This means that we can address and read each individual dataset inside of the array as needed. We can even save new data into any position inside of the array.

Arrays are distinguished by the brackets. Here's an example of an array of integers:

```
intarrayName[]={ integerOne, integerTwo, integerThree,...};
```

Arrays can be prefilled with data, which are separated by commas and contained between curly brackets {}. Arrays can be filled with data of the specified datatype, in this case integers.

In this example sketch, we have two arrays, melody[ ] and noteDurations[ ].

In the code example above, we are using a for-loop to iterate through each position of the two arrays using the variable "thisNote." As our for-loop iterates, we play each note inside of the melody[ ] array for the duration specified inside of the noteDurations[ ] array.

The sketch iterates through the eight elements of the two arrays to play each note for the specified duration using the `tone()` function.

The only modification that we have to make to this code example is to change the tone output pin to match our hardware setup. Change number eight in the `tone()` function to 19 as that is the PWM-capable pin our buzzer is connected to.

Make sure to make the same change for `noTone(19)`;

The last thing to note here is the additional delay between notes. This delay prevents the notes from being played immediately after the previous note ends. The comment reads:

```
// the note's duration + 30% seems to work well.
```

This suggests that the 30% value came about as a result of a series of trial-and-error experiments. What we want to point out here is that while for the most part everything we do is scientific and follows structured methodology, sometimes there are certain elements that you just have to tinker and play around with before you get something that "works." We are in the business of making and experimenting with new stuff, so it is always good to keep adjusting and trying things to get the result that you like. It is entirely possible that the 30% value in delay works well for the original developer's ears, but it might be completely off to you. So, if you think you can make it better, feel free to change, to adjust, to add in your own interpretation of what the sketch should be. Any change, any improvements, no matter how small, could mean new sketches, new features, new products, and new innovation.

If you're confident with your sketch, let's give it a go. It should be a melody that you have heard before!

Now that we are familiar with how we can use arrays to create a song, let's see what else we can do. To end this project, we challenge you to create your own LaunchPad-generated melody!

Challenges:

- Try to search for music notes on the internet to identify the appropriate notes and timing to play the song of your choice!
- Some suggested tunes are "Happy Birthday" or "Twinkle Twinkle Little Star."
- Go to the http://booksite.elsevier.com/9780124115880 to check out a few tunes that we have prepared for your enjoyment.

9.6 PROJECT 19: MUSICAL INSTRUMENT—POTENTIOMETER TO BUZZER

For this project, we want to use an analog input signal of the potentiometer to control the frequency of the buzzer output. This allows you to tune the pitch by dialing the potentiometer.

Ingredients:

1. LaunchPad
2. Educational BoosterPack
   OR

Breadboard, piezo buzzer (2.048 kHz), 10,000 Ω potentiometer, 100 Ω resistor

For those using the Educational BoosterPack, simply plug it into your LaunchPad as it includes the required potentiometer and buzzer. Again, be sure to remove the LED jumpers at P1.0 and P1.6 on LaunchPad. Also be sure to place the TXD and RXD jumpers appropriately if you want to enable simple debugging on the LaunchPad as well. Now you can plug in the Educational BoosterPack to your LaunchPad. Be sure to populate the jumper in the "POT" position on jumper J5 on the BoosterPack to ensure the potentiometer is connected to the analog input channel A3 on the MSP430G2553 device.

For the breadboard fans out there, here's how you can wire up the components on your own.

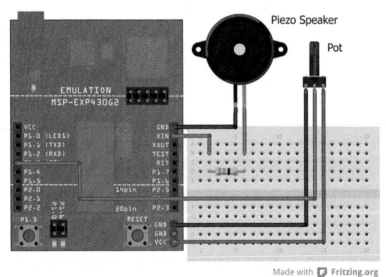

Made with 🟠 Fritzing.org

Luckily for us, there is also an existing example for this sketch inside of Energia.

Let's open up the example sketch in Energia from the following path:

File > Examples > 2. Digital > tonePitchFollower

Before digesting the code, let's first update the pin numbers to match our hardware setup.

1. Instead of using A0 as our analog channel, we are using A3. Which means `analogRead(A0)` needs to be changed to `analogRead(A3)`.
2. Our buzzer is located on pin 19, which means `tone(8, thisPitch, 10);` becomes `tone(19, thisPitch, 10);`

After making those modifications, your final code should look like this.

```
void setup() {
    // initialize serial communications (for debugging only):
    Serial.begin (9600);
}

void loop() {
    // read the sensor:
    int sensorReading = analogRead(A3);
    // print the sensor reading so you know its range
    Serial.println (sensorReading);
    // map the pitch to the range of the analog input.
    // change the minimum and maximum input numbers below
    // depending on the range your sensor's giving:
    int thisPitch = map(sensorReading, 400, 1000, 100, 1000);

    // play the pitch:
    tone(19, thisPitch, 10);
}
```

Note that this sketch sets up the Serial interface for debugging purpose. Therefore, in case you want to see the analog reading coming in, you can open up the Serial Monitor found under the Magnifying Glass to see the analog reading transmitted back to the PC.

The second thing to note in this sketch is that it does not directly use the sensorReading as the pitch for the tone generation. Instead, the analog reading is mapped to a pitch using the map() function.

```
int thisPitch = map(sensorReading,400,1000,100,1000);
```

We are using the map() function to take the potentiometer analog reading which should be in the range of [400, 1000], then scale it linearly within the range of [100, 1000]. The output of this map() function is then stored in a variable called "thisPitch." Mapping the input to the output scale essentially makes the entire input range usable, compatible, and relatable to the output range. This is simply a quick reminder since we used this function in the potentiometer project in the previous chapter.

And now you're set! Download the sketch to the LaunchPad and change the position of the potentiometer to see the change in pitch of the buzzer. This project is very similar to a previous project we did in Chapter 8 where we used a potentiometer to change the intensity of an LED. Instead of changing the duty cycle of a PWM signal that is tied to an LED, we are doing something a little bit different in this project. We are taking the analog signal from the potentiometer to change the frequency of the square wave that is tied to the piezo buzzer.

9.7 PROJECT 20: MUSICAL INSTRUMENT PART 2—ACCELEROMETER TO BUZZER

This project is a modification of Project 19. To reinforce the idea of signal chain and transformation, we will swap out the potentiometer to use a different analog input to affect the output noise being generated. For this project, we want to use an analog input signal of the accelerometer to control the frequency of the buzzer output. This allows you to change the output pitch by simple moving or tilting the LaunchPad.

Ingredients:

1. LaunchPad
2. Educational BoosterPack
   OR
   Breadboard, piezo buzzer (2.048 kHz), ADXL335 breakout board, 100 Ω resistor

Let's get the hardware set up. If you are using the Educational BoosterPack, make sure to place the jumpers on the BoosterPack to the ACC Y and Z positions. Once the jumpers are placed properly, plug it into your LaunchPad. As usual, remove the P1.0 and P1.6 jumpers that connect the red and green LEDs. If you want to be able to debug, make sure to place the RXD and TXD jumpers in the correct positions on the LaunchPad.

For breadboard users, here are the wiring instructions.

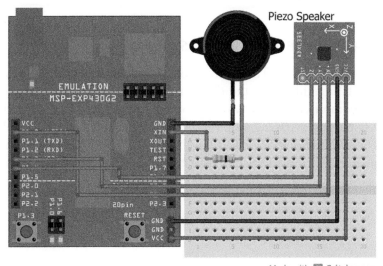

Made with 🅵 Fritzing.org

As we are simply changing the analog input, we can reuse the entire sketch from the previous project. The only change in the sketch is the input channel for the `analogRead()` function.

We can choose any of the X, Y, or Z channel to control the pitch of the buzzer.

- X = A0
- Y = A3
- Z = A4

We'll leave it up to you which channel of the accelerometer to use as the input signal! However, if you are using the Educational BoosterPack, make sure that you place the jumpers in the ACC_Y or ACC_Z positions if you are using those channels.

And that's pretty much it, go try out your new sketch. As you experiment with the sketch, if you are not satisfied with the result, you can come back and adjust different parts of the sketch such as the accelerometer channel and the parameters of the map function. Don't underestimate the power of trial and error. This process is important to improve the overall user experience for many applications.

To help find out what the best parameters would be for your map function, don't forget about our ability to debug. This will allow us to see what kind of accelerometer data we are seeing, and we can use that insight to pass in more appropriate values into our map function.

9.8 WRAPPING UP THE DIGITAL STREAM

And there you have it. The digital world isn't quite as black and white as we first thought! Sure, the data that we use to describe the digital world is comprised of just two numbers. However, by twiddling these bits and pins in meaningful patterns and speeds, these two numbers can help us portray more information beyond black and white. In the next chapter, we'll explore this further and learn how our LaunchPad can communicate with other external devices through digital communication schemes.

CHAPTER 10

The Languages of LaunchPad

As you learned in Chapter 9, it's pretty amazing what we can do with just a few 1s and 0s. In this chapter, we'll learn about even more things that we can do within a binary world. But first, let's take a look at your LaunchPad.

10.1 LAUNCHPAD—A SOCIAL BUTTERFLY

It's red, shiny and as we've learned in the last few chapters, it's got quite a lot going for it! While we can accomplish a lot of things with the LaunchPad by itself, it can be even more interesting when we introduce our LaunchPad to brand new friends. Yes, it's sad but our LaunchPad can be lonely, especially if we want to create more complex projects or systems. But before we can introduce our LaunchPad to other digital components, we need to find a way for our LaunchPad to both "speak" and "hear." Just like the way we communicate with others, our LaunchPad also needs a standardized language and interface. To communicate with other friends in a binary world, there are many languages, or standardized communication protocols that our LaunchPad can learn.

In this chapter, we're going to talk about some of these languages and different ways our LaunchPad can socialize with external systems. These languages and interfaces allow our LaunchPad to communicate with things such as your computer, digital components such as LCDs or sensors, or maybe even other LaunchPads.

10.2 A MULTILINGUAL LAUNCHPAD

While our LaunchPad's vocabulary may only consist of 1s and 0s, it's able to speak these bits in ways that other components can understand. The microcontroller on our LaunchPad is actually multilingual. It can speak different types of languages or communication protocols such as UART, SPI, or I2C. While these funny acronyms may not mean much yet, it means absolutely everything to communicating digital modules.

These three protocols fully define how various digital components should communicate to one another, much like we have rules in our own human language. By simply toggling 1s and 0s in an appropriate manner, our LaunchPad can talk to its digital friends! And as long as our LaunchPad can recognize what an incoming stream of 1s and 0s mean, our LaunchPad can finally hear and hold a conversation!

10.3 PROJECT 21: LAUNCHPAD, MEET COMPUTER, COMPUTER, MEET LAUNCHPAD (WITH UART)

To learn more about how our LaunchPad can digitally communicate with other modules, let's introduce our LaunchPad to its very first friend—the computer! In this case, we're going to jump back into Energia to teach our LaunchPad how to speak up!

Ingredients:

1. LaunchPad

The hardware setup for this project is simple. Simply plug in your LaunchPad to the computer. As we are going to communicate with the computer, be sure to place the RXD and TXD jumpers in the appropriate position on the LaunchPad. Once your hardware is set up, follow the steps below.

1. Type in the code below into an empty sketch inside of Energia.

```
void setup(){
    Serial.begin(9600); // initialize UART communication @ 9600 bps
}

void loop(){
    Serial.println("Hello computer!")
    Serial.println("Nice to meet you!");
    delay(1000);
}
```

2. Once you have it all typed up, compile and run the code.
3. Now that your LaunchPad has been reprogrammed, we can click on the Magnifying Glass that is on the top-right corner of the Energia window. You might remember this little guy from our Debug lesson in Chapter 8. If you get an error, be sure to check if you set up Energia to check the right serial port.

4. Now that we're in the Serial Monitor, what do you see? At this
   point, we should see the message that we typed into our sketch.

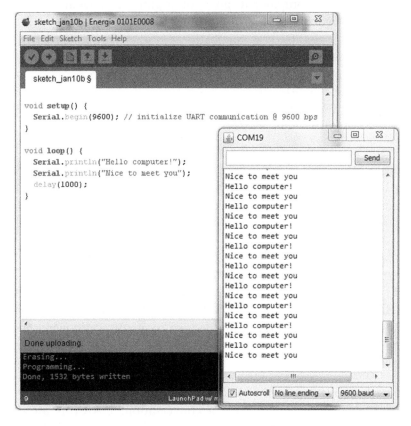

Congratulations! Your LaunchPad has broken the ice with the computer. Now that our LaunchPad has learned how to speak, we can use
this new capability for many things.

We actually have used the Serial Monitor several times already to
introduce simple debugging. While we didn't know it then, we were

actually using a UART protocol to send diagnostic information to the computer. UART provides an easy way for microcontroller developers to debug code, or find mistakes in the sketch, by taking a peek at the inside of the microcontroller brain.

UART transmission has many use-cases:

- Send commands to the computer.
- Provide status updates to the computer.
- Read out variable values to the computer.
- Debug your code by sending diagnostic information to the Serial Monitor on the computer to give you a glimpse of what is happening inside of your microcontroller brain.
- Also, UART enables communication to other components beyond the computer. We can use it to talk to other LaunchPads or other UART-capable devices.

Now that our LaunchPad has communicated with the computer, let's take a closer look at the code.

In our `setup()` function, the only thing we did was initialize the communication protocol inside of our LaunchPad. In this case, we're using UART, which is one of the serial communication languages that our LaunchPad is capable of speaking.

10.4 "WHAT IS YOU-ART?"

UART stands for *Universal Asynchronous Receiver/Transmitter*. Yeah, it's definitely a mouthful, a mouthful of 1s and 0s. As its name implies, UART is a universal standard that is used by many components inside of many electronics. The second thing to recognize is the word "asynchronous."

Asynchronous means the lack of synchronized timing between the involved modules trying to have a conversation. UART is an asynchronous language in the binary world, which means that before any digital friends can start talking to each other in the UART language, they first have to agree upon a consistent speed. Because UART is asynchronous, the first thing we have to do in our `setup()` function of our example code above is set a consistent speed that the computer is expecting. In this case, the Serial Monitor inside of Energia is expecting UART communication at 9600 bps. This speed is known as the "Baud Rate."

We can see that the Serial Monitor is expecting 9600 bps because of the setting that is defaulted on the Serial Monitor shown below:

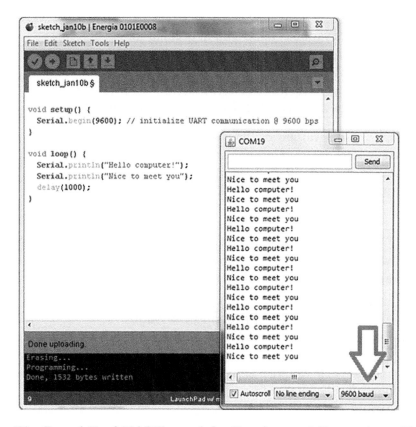

The R and T of UART stand for Receiver and Transmitter. This means exactly what it says. UART is a bi-directional communication protocol. However, in this example, our LaunchPad was only able to practise the "T" part of UART by transmitting those two messages to the computer.

So what else can this Serial.println() function do for us?

The Serial.println() function is capable of transmitting various types of data over UART.

• Strings of text: Serial.println("Hello computer!")

- Individual characters: `Serial.println(' A' )`
- Numbers: `Serial.println(125)`
- Variable values: `Serial.println(variableName);`
  - This function also has the ability to display data in different number formats including binary, decimal, or hexadecimal.
  - We can do this by passing in another parameter
    - `Serial.println(variableName, DEC);`
      // prints variableName data in decimal format
    - `Serial.println(variableName, BIN);`
      // prints variableName data in binary format
    - `Serial.println(variableName, HEX);`
      // prints variableName data in hexadecimal format

There's also a variant of the `Serial.println()` function call. As you noticed in our example above, there was a line feed between each transmitted message. If that line feed is not desired, we can use the following function instead:

```
Serial.print();
```

Now that our LaunchPad knows how to speak over UART, let's teach our LaunchPad how to listen over UART.

10.4.1 The "R" of UART—Receiving Messages
As we learned, UART is capable of both Receiving and Transmitting serial digital data. To learn about how our LaunchPad can listen to its digital friends that can also communicate over UART, let's play with the following code inside of Energia.

10.5 PROJECT 22: ECHO! ECHO! ECHO! ECHO!

In this project, we'll be doing bi-directional communication between a LaunchPad and a computer over UART.

Ingredients:

1. LaunchPad

The hardware setup is easy for this project. Simply plug your LaunchPad to your computer. Also, because we are using UART to communicate with the computer, be sure to properly place the RXD and TXD jumpers on the LaunchPad in the appropriate positions. Once the hardware is setup, jump to Energia and type the following code.

```
int incomingData = 0;  // this will catch the incoming UART data

void setup(){
    Serial.begin(9600); // initialize UART communication @ 9600 bps
}

void loop(){
    if (Serial.available() > 0){      // check if any serial data is
                                      // available to be read
        incomingData = Serial.read();      // if so, read it and
                                           // store it
        Serial.print("incomingData = ")    // send it to computer
        Serial.println(incomingData, HEX);
    }
}
```

Before we execute the code above, let's take a closer look at what we're trying to do.

First, we setup our UART transmission and configure the UART speed to be at 9600 bps. Remember, UART is asynchronous, meaning that we need to *explicitly* make sure that both speaker and listener are configured at the same speed.

One way to think of this are two piano players. Have you ever heard the song "Heart and Soul" before? While the name may not be familiar to you, you will probably recognize it upon hearing it. It's a simple song that has two clear parts—a bass and a treble component. Each part can be played by a different piano player. So what does this have to do with UART?

Well, in order for the two piano players to sound good together as they play their duet, they have to be synchronized and aligned to the same tempo. If one piano player is playing their part at 140 bpm and the other player is jamming away at 200 bpm, the song just wouldn't work! This is very similar to UART—both speaker and listener have to be in sync.

In our loop() function, we introduce a new function called Serial.available(). This function checks to see if any data is available to be read. If data is available, it will return a value, which indicates how many datasets are available to be read by our LaunchPad.

The first thing we do is use this new function to check if data is available. If data is available to be read, we use another function called `Serial`.read() to actually read out the next available data. The convenient thing about this `Serial`.read() function is that once it reads the next dataset, it removes it from the queue. This means that as long as no new data comes in, the next time we call `Serial`. available(), it should notify us that there is one less dataset in the queue.

This is what we call an FIFO queue. FIFO stands for "First In, First Out." So as data is received by our LaunchPad, it is queued in a line. Each time we call the `Serial`.read() function, it pulls out and reads the datasets one at a time in the order that they came into the queue.

Continuing down our code, we see that once we read in a dataset using our `Serial`.read() function, we use a `Serial`.print() and `Serial`.println() function to send the received dataset back to our computer.

We use `Serial`.print("incomingData = "); as a fixed string with no line feed, then we follow it up with `Serial`.println(incomingData, HEX); which prints out the value held inside our insideData variable and prints it out in a hexadecimal format. And as we used `Serial`.println(), we will get the automatic new line feed, which should help make our data easier to read inside of the Serial Monitor.

And that's it! We continue to loop here continuously checking if any data is available in the queue. If so, we read each dataset one-at-a-time and send it back to the computer with our print functions.

In a way, we could summarize this simple code example as an "echo." Our LaunchPad sketch simply echoes back to the computer the data that it receives from the computer.

Great—now that we have dissected this code, let's run it! Verify and Download your code.

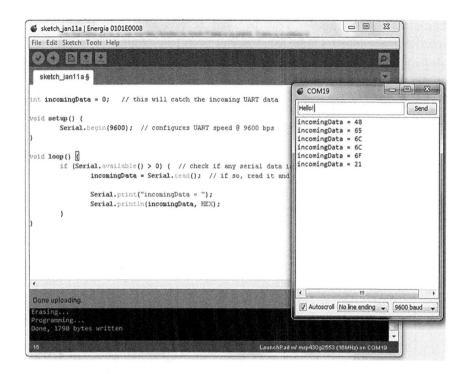

We will again use our trusty Serial Monitor, but this time we will use it to send data from the computer to our LaunchPad through UART. At the top of the Serial Monitor window, we can see there is an empty field and a "Send" button. It is through this field where we can send data to the LaunchPad via UART.

To test this out, let's send the message "Hello!" to our LaunchPad kit. Once we hit "send," we see that our LaunchPad immediately "echoes" back six messages over UART to the computer.

What exactly happened here? Well, our Serial Monitor sent our "Hello!" message one character at a time. In return, our LaunchPad sketch then received each character and echoed it back to the computer one-at-a-time.

10.6 TRANSLATING LAUNCHPAD—SPEAK WITH ASCII

What we will notice, however, is that our LaunchPad is echoing back hexadecimal values, just like we asked it to in our LaunchPad sketch. The reason we passed back hexadecimal values is because this is a convenient way to show how our LaunchPad understands English characters and symbols. This is what is typically called ASCII or American Standard Code for Information Interchange. Because our LaunchPad brain can only understand 1s and 0s, ASCII is a way for the microcontroller to handle characters and symbols.

Here's the ASCII table.

Dec	Hex	Symbol	Dec	Hex	Symbol	Dec	Hex	Symbol	Dec	Hex	Symbol
0	00	NULL	32	20		64	40	@	96	60	`
1	01	SOH	33	21	!	65	41	A	97	61	a
2	02	STX	34	22	"	66	42	B	98	62	b
3	03	ETX	35	23	#	67	43	C	99	63	c
4	04	EOT	36	24	$	68	44	D	100	64	d
5	05	ENQ	37	25	%	69	45	E	101	65	e
6	06	ACK	38	26	&	70	46	F	102	66	f
7	07	BEL	39	27	'	71	47	G	103	67	g
8	08	BS	40	28	(	72	48	H	104	68	h
9	09	HT	41	29	)	73	49	I	105	69	i
10	0A	LF	42	2A	*	74	4A	J	106	6A	j
11	0B	VT	43	2B	+	75	4B	K	107	6B	k
12	0C	FF	44	2C	,	76	4C	L	108	6C	l
13	0D	CR	45	2D	-	77	4D	M	109	6D	m
14	0E	SO	46	2E	.	78	4E	N	110	6E	n
15	0F	SI	47	2F	/	79	4F	O	111	6F	o
16	10	DLE	48	30	0	80	50	P	112	70	p
17	11	DC1	49	31	1	81	51	Q	113	71	q
18	12	DC2	50	32	2	82	52	R	114	72	r
19	13	DC3	51	33	3	83	53	S	115	73	s
20	14	DC4	52	34	4	84	54	T	116	74	t
21	15	NAK	53	35	5	85	55	U	117	75	u
22	16	SYN	54	36	6	86	56	V	118	76	v
23	17	ETB	55	37	7	87	57	W	119	77	w
24	18	CAN	56	38	8	88	58	X	120	78	x
25	19	EM	57	39	9	89	59	Y	121	79	y
26	1A	SUB	58	3A	:	90	5A	Z	122	7A	z
27	1B	ESC	59	3B	;	91	5B	[	123	7B	{
28	1C	FS	60	3C	<	92	5C	\	124	7C		
29	1D	GS	61	3D	=	93	5D	]	125	7D	}
30	1E	RS	62	3E	>	94	5E	^	126	7E	~
31	1F	US	63	3F	?	95	5F	_	127	7F	DEL

Using the table above, we can see that the letter "H" has an ASCII value of 0x48 in hexadecimal. This is the same hex value that we see in our Serial Monitor as the first echoed character.

Now that it's able to establish two-way communication to a computer, our LaunchPad is well on its way to become a social butterfly. Now that we've accomplished and exercised both the "R" and "T" of UART, let's take a closer look at what's happening at a hardware level.

As we mentioned in the beginning of this chapter, communication between LaunchPad and its digital friends can occur when we have standardized serial communication protocol and physical interface. UART provides both of those pieces.

10.7 THE HARDWARE BEHIND UART

Let's first look at the physical interface required for UART communication. UART is a two-wire protocol and consists of a TXD (transmit) and an RXD (Receive) line. We know that the "A" in UART stands for asynchronous because the two systems trying to talk to each other do not share a clock line. In synchronous communication, a clock signal is shared between the systems that are trying to communicate with each other. However, UART is asynchronous so no clock signal is required. Instead, we have to configure the UART on all communicating systems at the same baud rate.

Also, our LaunchPad communicates to the computer using a method called, "Full Duplex." This means that we can transmit and receive at the same time. UART can also be implemented in a "Half Duplex" method, which means that a system can only send or receive at any given time. In the case of Half Duplex, the RX and TX line can be shared, meaning that it can be implemented with just a single wire. This is good for systems that do not have very many pins available.

However, in our LaunchPad we are using the Full Duplex method, which requires two wires and enables RX and TX to happen simultaneously. The image below shows the hardware implementation of a Full or Half Duplex setup.

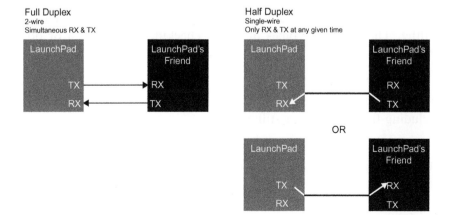

Now that we know how UART is set up physically between various systems, let's examine the serial communication protocol that UART follows. As we know, our LaunchPad and its digital friends speak a digital language made up of 1s and 0s. To make sense of the 1s and 0s that are being transmitted and received, UART specifies how these data bits can be sent between communicating systems.

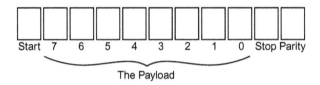

First, we have a "Start" bit. This bit signifies the start of a UART transmission.

We follow the Start bit with a byte of data (or 8 bits). These 8 bits can be any series of 1s and 0s and this makes up your data "payload."

After the payload of 8 bits is sent, UART follows it up with a "Stop" bit signifying the end of the transmission. A payload can be of varying lengths; however, in our examples, we will use a payload of 8 bits.

The Start and Stop bits frame a single transaction of our UART transmission.

A "Parity" bit can also be included in a UART transmission and typically follows the Stop bit. This parity bit is usually used as a form of error checking. Just like when you speak on the telephone, sometimes the audio quality is choppy and your friend cuts in and out or sounds garbled, UART and other forms of serial communication can also be corrupted. This parity bit is used as a simple form of error checking to see if a message that is received via UART is corrupted or healthy. There are two types of parity checking—either even or odd. Both cases help to check for single bit errors.

In Even Parity checking, our parity bit is set as a 0 if an even number of 1s are available in the payload. Thus, the total number of 1s (including the parity bit) is still even.

If an odd number of 1s are inside of your payload, the parity bit is set to a 1. Again, this way the total number of 1s (including the parity bit) is made even.

In Even Parity, the number of 1s in your payload + the parity bit should always be even. So, if the receiving end of the UART transmission counts the total number of 1s (between payload and parity bit) and finds an odd number, it knows that this is a corrupted and erroneous data transmission. Because we only know that the number of 1s are incorrect, we cannot actually fix or correct the data transmission. We can only scrap it.

In Odd Parity, it is very similar except that we use the parity bit to make the total number of 1s in the payload + parity bit should always equal an odd number of 1s. This way, if the receiving component counts an even number of 1s, it knows that this is a corrupted transaction.

10.8 PROJECT 23: CONTROLLING LAUNCHPAD WITH COMMANDS FROM COMPUTER

You're well on your way to become a UART master! Now that we know the basics of UART, let's make things interesting by learning how to control the red and green LEDs on our LaunchPad through UART commands from the computer!

Ingredients:

1. LaunchPad

The hardware setup is simple as we will be simply using the onboard red and green LEDs on the LaunchPad itself. Be sure that the jumpers are connected to P1.0 and P1.6 allowing the LEDs to be electrically attached to the pins.

In this project, we will use the Serial Monitor in Energia to toggle either the red or the green LED on the LaunchPad. We will do that by sending the following commands:

- Sending "R" or "r" = toggle the red LED on/off
- Sending "G" or "g" = toggle the green LED on/off

Here's the code that we can use to make this happen. Type it into Energia.

```
int incomingData = 0; // this will catch the incoming UART data
int greenState = LOW; // this will remember the state of the green LED
                      // (HIGH or LOW)
int redState = LOW;   // this will remember the state of the red LED
                      // (HIGH or LOW)

void setup() {
Serial.begin(9600);  // configures UART speed @ 9600 bps
pinMode(GREEN_LED, OUTPUT); // these functions look familiar!
pinMode(RED_LED, OUTPUT); // Need to configure the pins connected to our
                          // LEDs as outputs
}

void loop() {
if (Serial.available() > 0) {  // check if any serial data is available
                               // to be read
    incomingData = Serial.read();  // if so, read it and store it

    Serial.print("incomingData = ");
    Serial.println(incomingData, DEC);  // echo received data back to
                                         // computer

  //Check to see if incoming character is a 'r', 'R', 'g' or 'G'
  if(incomingData == 82 || incomingData == 114){ // 'R' = 82 in decimal
                                                  //   per ASCII key
                                                  // 'r' = 114 in decimal
                                                  //   per ASCII key
    togglePin(RED_LED); // custom function we created to toggle the
                        // LED state
    }
else if (incomingData == 71 || incomingData == 103){ // 'G' = 71 in
                                                      // decimal per ASCII key
                                                      // 'g' = 103 in decimal
                                                      //   per ASCII key
         togglePin(GREEN_LED);  // custom function we created to
                                // toggle the LED state
    }
  }
}

/*Custom function that we have created, which toggles the appropriate
pin based on previous state of the desired LED. We pass in the desired
LED we want to toggle as a parameter*/

void togglePin(int LED){
    if(LED == RED_LED){ // we want to toggle the red LED
        if(redState == LOW){ // if low, toggle high
            digitalWrite(RED_LED, HIGH);
            redState = HIGH;
        }
```

```
else if(redState == HIGH){  // if high, toggle low
        digitalWrite(RED_LED, LOW);
        redState = LOW;
        }
    }
    else if(LED == GREEN_LED){  // we want to toggle the green LED
        if(greenState == LOW){  // if low, toggle high
        digitalWrite(GREEN_LED, HIGH);
        greenState = HIGH;
    }
    else if(greenState == HIGH){  // if high, toggle low
        digitalWrite(GREEN_LED, LOW);
        greenState = LOW;
    }
    }
}
```

Once you've got the code all set up, let's run it through its paces! Compile, download and run the code on our LaunchPad. Does it do what we want it to do?

In this project, we run through a few concepts here. Use the comments in the code above to understand what is happening. We also leveraged our newly-acquired knowledge thanks to the ASCII table to help us decipher between the characters and the symbols we are more familiar with as humans and the 1s and 0s that our LaunchPad can understand. Thanks to our understanding of ASCII, we can tell if the user is typing in a capital or lower case "r" or "g."

10.9 PROJECT 24: THE COLOR PICKER

In this exercise, we're going to change the color of an RGB LED using commands sent from the computer by passing in the desired RGB values from the computer. We can reuse the same schematic that we used in the previous chapter. Or if you have the Educational BoosterPack, you can use that too!

Ingredients:

1. LaunchPad
2. Educational BoosterPack
   OR
   Breadboard, RGB LED, $1 \times 150 \ \Omega$ resistor, $2 \times 100 \ \Omega$ resistor

For those with the Educational BoosterPack, simply plug it into your LaunchPad. However, be sure to place the TXD and RXD jumpers appropriately to enable UART communication. Also be sure to remove the LED jumpers at P1.0 and P1.6.

For those who want to breadboard your own solution, use the diagram below for the hardware setup.

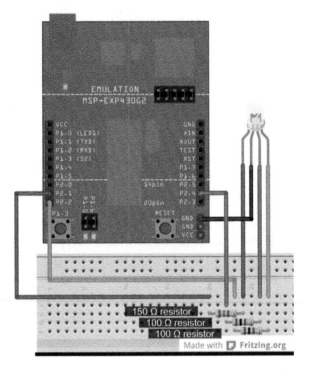

In this exercise, we'll learn how to parse through the incoming data from the computer. We want to create the desired color on the RGB LED based on a command that will consist of three values. We will control color using its RGB color code in the following format: R_color_code, G_color_code, B_color_code. Each color code is a value between 0 and 255. By changing the brightness of each LED inside of our RGB LED, we can create a wide array of colors! We will pass in the desired RGB color code over UART and will separate each RGB intensity with commas.

Here are a few example commands our sketch should recognize over UART:

- 255,0,0 → Red
- 0,255,0 → Green
- 0,0,255 → Blue
- 255, 0,255 → Purple

To help us digest the incoming UART data, we will be using a concept we introduced earlier in the book—the array. As a reminder, arrays are used to carry a set of data, and it is capable of holding various data types including integers, characters, and more.

An array needs to be declared just like any other variable. Here's an example of how we can initialize an array called "incomingMessage":

```
int incomingMessage[12];
```

incomingMessage has been declared as an array of integers, and it has been initialized to reserve 12 slots. In this case, the 12 slots inside of our array have not been initialized. Just like with any other variable, we also have the option of initializing our array with default values. We can do this like so:

```
int incomingMessage[12]={1;2;3;4;5;6;7;8;9;10;11;12};
```

In this project, we'll be using an array to hold the ASCII code for the incoming data that is flowing in through UART. Once we receive a full RGB command over UART, we'll be using the array to separate the data associated to each color—red, green, or blue. Because our data will be separated by commas (,), we will be able to easily decipher which elements within our array corresponds to each color.

Arrays are useful to carry a particular dataset. We'll be using an array similar to the example above in this project to carry our incoming RGB command that our LaunchPad will receive from the computer. Within this project, we'll also introduce how to iteratively go through each element of an array with the help of for-loops and an index variable.

Once our LaunchPad receives the RGB command over UART and deciphers it, we will be changing the intensity of each RGB LED using a function that we should already be familiar with—analogWrite().

But, before we execute the code, there is a setting we'll have to flip inside of Energia's Serial Monitor. Within the Serial Monitor, we can tell it to send or discard the ASCII code associated to a user pressing the "Enter" button. As you will see in the code below, we need to know when a user presses "Enter" so we know that the user has sent out a complete RGB command over UART.

To tell the Energia Serial Monitor to also send the ASCII value for "Enter" (10 in decimal), we have to select the "Newline" option shown in the screenshot below.

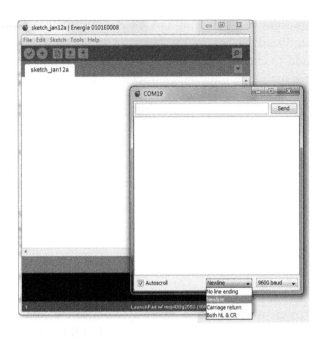

So, when a user sends an **RGB** code of "255,255,255" over **UART** using the Serial Monitor, our LaunchPad will receive the following ASCII code for each character:

ASCII Code (Decimal)	Character
50	2
53	5
53	5
44	,
50	2
53	5
53	5
44	,
50	2
53	5
53	5
10	NewLine

Well, let's get going shall we!

Also, before executing the code below, remember that this **RGB** LED can get quite bright! Try using lower **RGB** values or place a piece of paper on top of the LED to diffuse the light.

Tip

This is a long piece of code, so if you don't feel like retyping 3 pages worth of code, you can go to the http://booksite.elsevier.com/9780124115880 and download the sketch in its entirety. We have placed the code in the book here for your reading convenience.

```
/***************************************************
Send the RGB value of desired color.
Send with this format: R,G,B
Where R, G or B can be a value between 0 to 255
Examples:
255,0,0 --> Red
0,255,0 --> Green
0,0,255 --> Blue
255, 0, 255 --> Purple
***************************************************/

int incomingData = 0; // this will catch the incoming UART data

// Let's rename the pins attached to each channel of our RGB LED
int R = P2_1;
int G = P2_2;
int B = P2_4;

// this will be between [0,255]. It controls the duty cycle for the PWM
// on each LED
int red =0;
int green =0;
int blue =0;

// this array will hold incoming data
int incomingMessage[12];

// amount of data received until user presses "Enter"
int dataCount = 0;

// we'll use this to parse through the incomingMessage array
int k = 0;

void setup() {
    Serial.begin(9600); // configures UART speed @ 9600 bps
    pinMode(R, OUTPUT); // set pin connected to R as output
    pinMode(G, OUTPUT); // set pin connected to G as output
    pinMode(B, OUTPUT); // set pin connected to B as output
}

void loop() {
/* First, we want to continue to read UART until we see that the user
pressed "Enter". We also want to keep count of how many characters we
received until the user pressed "Enter". Notice that if Serial.read()
returns -1, that means that the receiving buffer was empty and nothing
was available to read. We will use a while-loop to continually read
incoming UART characters and will fill in the array incomingMessage[]
until the user presses "Enter" */
    dataCount = 0; // initialize our received data counter
    incomingData = Serial.read(); // read incoming data
```

```
    /* continually read incoming data until user presses "Enter"
       (Enter ASCII = 10)*/
    while(incomingData !=10){
          if(incomingData != -1){ // Serial.read() returns a -1 if
                                  // there is nothing to read.
              incomingMessage[dataCount] = incomingData;
                                    // if valid data,add to array
              dataCount++;          // increment data counter
          }
          incomingData = Serial.read();
                          // read next, then repeat while loop

    }
/* Awesome! We only break out of the while-loop above when the user
presses "Enter". Now that the array incomingMessage[] holds the complete
RGB command, we need to convert the characters into number form. For
example '2' '5' '5' needs to be converted to the number 255 so that we
can use it as an input to the analogWrite() function */

    k=0; // Initialize index to parse through the incomingMessage[]
         // array we populated above.

    /* Parse through array until we see first comma(,). This should be
    our Red data*/
    red = incomingMessage[k] - 48;
    /*convert ASCII to decimal. Notice that ASCII code for a number is
    offset by 48 with the number that it represents.*/

    /* increment array index to read the next saved data */
    k++;
    /*continually parse through array until we see next comma (,)
    using while-loop */
    while(incomingMessage[k] != ','){
          /* Multiply previous digit by 10 to shift it a decimal place
          then add in the new converted digit to update the Ones
          place*/
          red = red*10 + (incomingMessage[k]-48);
          k++; // repeat while loop until next comma is seen.
    }

/* comma was found so we break out of the while loop. Increment to next
item in array and start decoding the green value */
    k++;

    /* to extract green data from our array, do the same as we did to
    find the red data.*/
    green = incomingMessage[k] - 48;
    k++; // increment array index
    while(incomingMessage[k] != ','){
          /* Multiply previous digit by 10 to shift it a decimal place
          then add in the new converted digit to update the Ones
          place*/
          green = green*10 + (incomingMessage[k]-48);
          k++; // repeat while loop until next comma is seen.
    }
```

```
/* comma was found so we break out of the while loop. Increment to next
item in array and start decoding the blue value */
    k++;

    /* To get the blue data, it's a little different since we don't
    have another comma to find.Instead,we parse through our array
    until we reach the end of it using a for-loop.*/
    blue = incomingMessage[k] - 48; // convert ASCII to decimal
    k++; // increment array index
    for(k; k < dataCount; k++){ // continually parse through the rest
                                // of our array.
        blue = blue*10 + (incomingMessage[k]-48);
                                // convert to decimal
    }

/*now that we've converted the ASCII values to the values our LP can
understand, use analogWrite() to change the intensity of each color
channel.*/
    analogWrite(R,red);
    analogWrite(G,green);
    analogWrite(B,blue);

//echo back the color command received back to the computer
    Serial.print(red);
    Serial.print(',');
    Serial.print(green);
    Serial.print(',');
    Serial.println(blue);
}
```

Compile and download this code in Energia. Using the serial monitor, send in the R, G, B color codes to generate the appropriate color on the RGB LED. Use the comments that are embedded in the sketch above to get a better understanding of what's happening here.

We've learned quite a lot since we started this chapter. We've taken a look at the UART interface on a hardware standpoint, understanding its RX and TX lines. We've also learned how to use arrays to parse strings, ASCII and a few new function calls that we can add to our box o' tricks including `Serial.read()` and `Serial.available()`. You now also know the difference between Half and Full Duplex and can also explain how duelin' piano players have something in common with UART (asynchronous protocol and tempo)!

10.10 SERIAL PERIPHERAL INTERFACE

As we mentioned at the beginning of this chapter, our LaunchPad can speak to many different digital friends in various digital languages. The next language that we'll talk about in this chapter is called Serial Peripheral Interface, or SPI. Just like UART, SPI is a serial communication protocol that enables our LaunchPad talk to other external peripherals.

These external peripheral modules that we can communicate with using SPI include LCD displays (we'll do this in a later project), wireless modules, sensors, and again, even other LaunchPads. SPI is a little different compared to UART. For one, SPI is a *synchronous* communication protocol, which means that a clock signal is required and shared between the communicating modules. This is different from UART, which required us to explicitly set the same speed on both sides of the UART conversation. In SPI, however, both sides of the conversation are in-sync thanks to a shared clock signal that provides accurate and consistent timing between the speaker and the listener.

Also, SPI is a Full Duplex communication protocol, which means that our LaunchPad and its digital friends are capable of sending and receiving information at the same time. Another great thing about SPI is that it is a great solution for times when your LaunchPad is feeling especially sociable. The reason for that is because several modules can share the same communication lines, which enables our LaunchPad to talk to multiple modules for more complex conversations or systems.

Let's take a look at SPI and its hardware interface. The first thing to note is that SPI is a four-wire protocol.

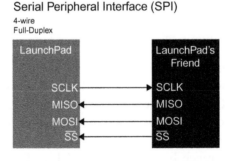

Serial Peripheral Interface (SPI)
4-wire
Full-Duplex

In this hardware setup, we can see there are four wires required for this type of communication. Let's go through them one at a time.

SCLK—This is a clock signal, which allows LaunchPad and its friends to stay in-sync for a successful conversation. Remember, SPI is a synchronous serial protocol!

MISO—MISO stands for Master-In, Slave-Out. In this example, our LaunchPad is the Master, which means that it is in control of this conversation. As a Full Duplex protocol, this line is dedicated for the sole purpose of enabling our LaunchPad to hear what its friends are saying (we have a separate line for communication in the other direction). As the Master, our LaunchPad also drives the SCLK signal as well to keep the conversation at the right pace.

MOSI—Yup, you guess it. This means Master-Out, Slave-In. This line is the other half of our Full Duplex interface, which enables our LaunchPad to talk to its friend(s).

$\overline{\text{SS}}$—This is our Slave Select line. As we mentioned, SPI is a great option for when our LaunchPad wants to talk to many friends. As the Master of this conversation, our LaunchPad uses a Slave Select line to choose which friend to talk to. You might notice that there is a bar over SS in our diagram. This is because Slave Select is an *active low* signal. This means that to address a particular slave, or friend, our LaunchPad should drive that specific line LOW, or set that pin to a 0.

Now that we can see how a one-on-one conversation is set up between LaunchPad and one of its friends, let's take a look at a more sociable event where our LaunchPad has multiple friends available for a chat.

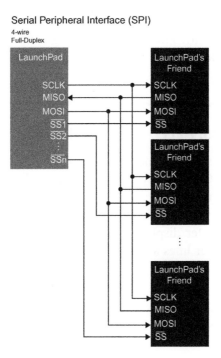

Serial Peripheral Interface (SPI)

Our once-lonely LaunchPad has sure come a long way! It's now the life of the party in this multislave system. In this scenario, our LaunchPad is again the Master of the conversation. Our LaunchPad drives the shared clock signal, which keeps all of the modules communicating at the right speed. Additionally, all of LaunchPad's friends are sharing the same MISO and MOSI lines, which enables a Full Duplex conversation with all of the modules. To ensure that the MISO and MOSI lines of communication don't get too muddied up, our LaunchPad takes full control of the conversation by using the various Slave Select lines. Each of LaunchPad's friends has its own Slave Select line, which can be connected to a specific pin on our LaunchPad. When LaunchPad wants to communicate to any one of its friends, the associated Slave Select line is activated. Another thing to note is that our LaunchPad follows great conversation etiquette because when it speaks to one of its digital friends, our LaunchPad will give that particular friend its full and undivided attention! To have a conversation with a specific friend, our LaunchPad must drive the particular pin that controls the specific Slave Select line LOW (remember this is an active low signal). And to ignore the other friends that might be trying to talk, our LaunchPad can drive the Slave Select line HIGH to all of the other modules to "silence" them.

Phew—we made it! Thanks for sticking with us this far as we took a deep look at the SPI interface. Now, let's do something cool with SPI!

10.11 PROJECT 25: HELLO WORLD WITH LCD

In this project, let's meet a new friend of our LaunchPad—a Liquid Crystal Display (LCD). Many LCDs are controlled over an SPI bus. Through SPI commands, we can control the LCD and tell it what type of messages to show on its display. In this example, we'll be playing with a 16×2 character display. This means that with this display, we have two 16-character lines at our disposal.

Ingredients:

1. LaunchPad
2. Educational BoosterPack
   OR

Breadboard and LCD (Part Number NHD-C0216CZ), 2 × 10,000Ω resistors, 1 × 15,000Ω resistor, 2 × 1 uF capacitors

For this project, we'll display "Hello World!" on one of the available 16-character lines. We will also display a counter that increments once per second on the second 16-character line of our LCD display. To do this, we will actually leverage a pre-existing *library*.

Define

Library

    Library is a collection of functions that can be imported into our LaunchPad's brain. Just like we can use various functions that are inherently available inside of Energia, or create our very own functions from scratch, another way of adding new capabilities to our LaunchPad is by importing a library.

If you recall from the previous chapter, a library is a collection of information and function that can be pulled into our sketch. In this case, because we want to talk to an LCD display over SPI, we can import an existing LCD library. This LCD library will provide us with many functions that we will need to display messages on the display.

One of the main benefits of using a library is that they provide an extra layer of abstraction away from lower level interfaces. For example, the LCD library that we will leverage for this exercise will allow us to focus on the functions of our LCD without requiring us to think too much about what is happening at the lower SPI layer. This abstraction is true for many libraries! Libraries are usually created to modularize functionality by grouping together particular actions, especially when related to modules external to our LaunchPad. These libraries allow developers to focus on the bigger picture and not have to worry about the bits and bytes happening underneath within our SPI layer!

So first, how do we import a library? Well, it's actually quite easy. We simply navigate to the following menu:

Sketch > Import Library...

As you can see in the screenshot, this menu inside of Energia presents to you various libraries that are available for us to import into our sketch. Before we import any libraries into our sketch, let's take a second to learn about how to get new libraries to pop up inside of this menu within Energia.

●●●────────────────────────────────

Tip

Libraries can be created by anyone! There is a large and growing user-base for LaunchPad, so who knows? Maybe one of your fellow LaunchPad enthusiasts have already created a library for an external module you are trying to interface with? In this case, your favorite search engine is your friend! The microcontroller development community is a friendly one that is usually very excited to help and share code examples and libraries.

As an example, we have a useful LCD library specifically created for the LCD on the Educational BoosterPack that we can use in this exercise that you can download and install from the web. You can find it at http://boosterpackdepot.com/wiki/index.php?title=Educational_BoosterPack under the Software section.

Once you have the library downloaded, copy and paste the LCD library into the "Libraries' folder inside of your Energia installation path. You can find it in the path below:

[Energia root folder] > hardware > msp430 > libraries

You may already have a library of the same name. If your computer asks you if you want to overwrite the existing library, go ahead! This library that we are importing is a new version that includes support for the LCD module that we are using for this exercise!

Great! Now that we have the latest version of the LCD library, let's import the new "LiquidCrystal" library into a blank sketch inside of Energia. Doing so will insert the following line into the top of your sketch.

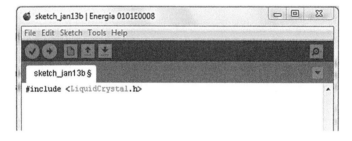

To make a particular library available to you within your sketch, all we have to do is to import it through the Energia top menu as shown above, or manually type that single line of code at the very top. We can also import multiple libraries into a sketch if more than one is needed for your particular project.

```
#include <libraryName1.h>
#include <libraryName2.h>
#include <libraryName3.h>
```

So now that we have our LiquidCrystal library imported into our sketch, let's get the hardware set up for this project.

If you have the Educational BoosterPack handy, plug it into your LaunchPad. We will be controlling the LCD display on the Educational BoosterPack in this project with the help of the LiquidCrystal library that we just imported. As usual, be sure to remove the LED jumpers P1.0 and P1.6. Also, place the RXD and TXD jumpers to enable UART back to the computer for debugging purposes.

If you don't have an Educational BoosterPack, have no fear! We can interface our LaunchPad to the LCD display using the breadboard layout below. The LCD we are using is the NHD-C0216CZ.

Here is a screen capture of the schematic of the board, followed by the fritzing-generated schematic.

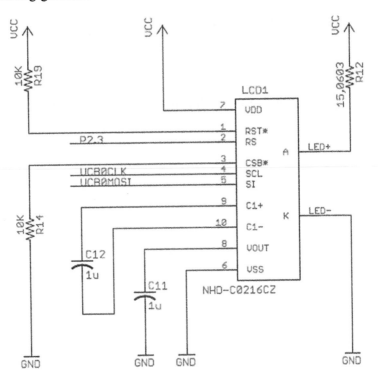

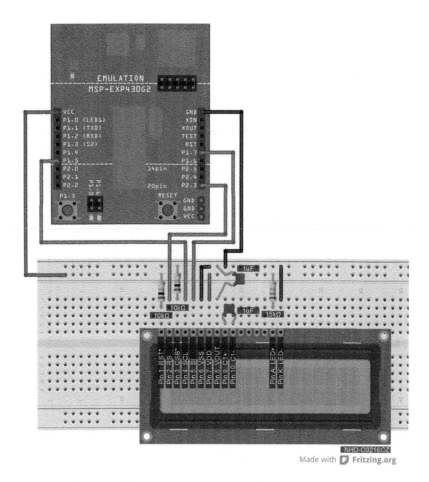

Made with **Fritzing.org**

In the diagram above, can you spot the SPI connections? As we only want to communicate to the LCD module, we really only need three lines for this SPI implementation: Slave Select, a communication line from LaunchPad to LCD and a shared clock line. The MSP430 microcontroller on our LaunchPad is the Master, while the LCD module acts as the Slave. One thing to note, however, is that we are only required to do MOSI (Master-Out-Slave-In) communication in this example. There is no need for the LCD to talk back to the LaunchPad in this particular application. So we can see that we actually only need three lines to enable the SPI communication for this example. We have our Slave Select line (CSB*), SPI Clock line (SCL), and MOSI line (SI). To simplify this example, we are simply connecting the Slave Select line directly to ground. Remember, because the Slave Select line is an active low signal, permanently tying it to ground (0 V) means that the LCD is always enabled for SPI communication.

Beyond the pins required for SPI communication, we also have a few other pins on the LCD that we have to connect to provide power to the LCD and control the LCD's backlight LED, among other things.

Now that we have our hardware setup, with either a breadboard or the Educational BoosterPack, let's get back to our code.

```
#include <SPI.h>
/* we have to import the SPI library as well since the LiquidCrystal
library needs it.*/

#include <LiquidCrystal.h>

int LCD_SS = 0; // We are not using the SS line in our hardware setup
int LCD_RS = P2_3; // We can see that the RS line of our LCD is tied to
                   // pin P2.3 on our LaunchPad.

LiquidCrystal lcd(LCD_SS, LCD_RS);
/* This line declares a LiquidCrystal type named "lcd".
   This declaration also expects us to tell it which pins
   our SS and RS lines are connected to */

void setup() {
    lcd.begin(16, 2);  // initialize LCD by specifying the dimensions
                       // of display (columns, rows)
    lcd.print("Hello World!"); // this line prints out the desired
                               // message that we want displayed
}

void loop() {
    lcd.setCursor(0, 1); // re-position the cursor for where you want
                         // to print next
    lcd.print(millis()/1000); // this function prints out the amount
                              // of seconds that have elapsed.
}
```

Go ahead and run the code above inside of Energia. If all goes well, you should see our "Hello World!" message as well as a counter.

To get this example up and running, we've introduced a few concepts and functions. Let's take a closer look at what's new here.

The first thing we notice is that we're leveraging a few libraries. At the top of our code are two libraries which we have imported called SPI and LiquidCrystal. These libraries are required to enable our SPI communication to our LCD module, and they provide a few useful functions that make it easier for us to accomplish our goals.

The next thing we do is initializing a few port pins. We need to tell the library which port pins we are using to connect the LCD to our LaunchPad. As you can see on the schematic of our hardware setup, the LCD_RS pin is connected to P2_3, which is what we configure in our code.

Next on our code is something that we haven't seen before. We are declaring a new LiquidCrystal variable type called "lcd." This unique LiquidCrystal variable type was made available by importing our LiquidCrystal library. Once we declare a new LiquidCrystal variable named "lcd" we can start using it!

The LiquidCrystal library provides us with a few nifty functions. Below are the three new functions that we can use thanks to this library:

```
lcd.begin(number_of_columns, number_of_rows);
/* This function initializes our lcd module and notifies the library the
"dimensions" of our display. In our case, the LCD has 16 colums and 2
rows, resulting in a total possibility of 32 individual characters or
symbols to be displayed at once.*/

lcd.print("your text here!")
/* This function displays the desired string on the LCD screen! This
function will print the desired message at the current location of the
cursor. By default, the cursor is located on the first row, first
column. It's important to remember that the LCD is only 16 characters-
wide! This means you need to ensure you don't pass in too many charac-
ters that the display cannot handle.
As you can see in this code example, we can pass in either a string such
as "Hello World" or a number value, which we used to display the elapsed
number of seconds. */

lcd.setCursor(desiredColumn,desiredRow);
/* This function places the cursor that you want the print function to
eventually start writing to. The first column on the display is column
#0, while the first row on the display is row #0.*/

lcd.clear();
/* This is another function that is available, which we did not use in
this lab, but it will prove useful in the upcoming lab. This function
doesn't require any parameters and it allows us to clear the screen. */
```

Those are the basic functions that are provided by the LiquidCrystal library. Now that we know how to display text and values on our fancy new SPI-driven LCD display, what else can we do?

Challenges:

One place to start is to try and incorporate this new LCD functionality into some of our older projects! Here are a few ideas:

- Display the ADC reading from the potentiometer scroll wheel.
- Display temperature in degrees Fahrenheit on the LCD screen.
- Display the RGB color code that we pass in over UART using the Energia Serial Monitor, while generating the appropriate color on the RGB LED.

We're well on our way to becoming fluent in the serial language of SPI!

10.12 PROJECT 26: A MAGIC ~~EIGHT-BALL~~ LAUNCHPAD

Create a magic LaunchPad that displays a random response on the LCD display when the LaunchPad is shaken.

Ingredients

1. LaunchPad
2. Educational BoosterPack
   OR
   Breadboard, ADXL335 three-axis accelerometer, 1×1 kΩ, 2×10 kΩ, 1×15 kΩ resistors, 4×0.1 uF, 1×4.7 uF, 2×1 uF capacitors

In this particular project, we'll be creating a fairly complex system that will require a three-axis accelerometer and the LCD. We will be combining a few concepts that we have recently learned about including analog signals from the accelerometer as well as SPI serial communication to an LCD module.

First, let's get the hardware set up. If you have the Educational BoosterPack, setup is quite easy. Simply plug in your Educational BoosterPack onto your LaunchPad and voila! Be sure that the two jumpers on the BoosterPack enable the Y and Z channels of the accelerometer to connect to the LaunchPad. Also be sure to remove the LED jumpers P1.0 and P1.6 on the LaunchPad. And lastly, position the TXD and RXD lines if you want to enable UART connection back to the computer.

For those who are Educational BoosterPack-less, we can recreate the required hardware using a breadboard, a few components, and some elbow grease! Below is a hardware diagram that shows how you can connect the accelerometer and the LCD module to the LaunchPad properly. Again, we've provided snapshots of the actual schematic as well as a diagram created using fritzing.

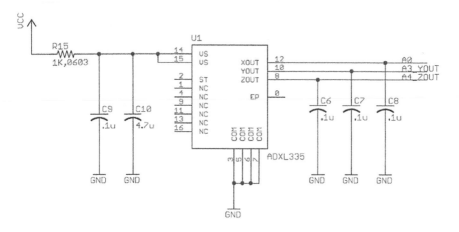

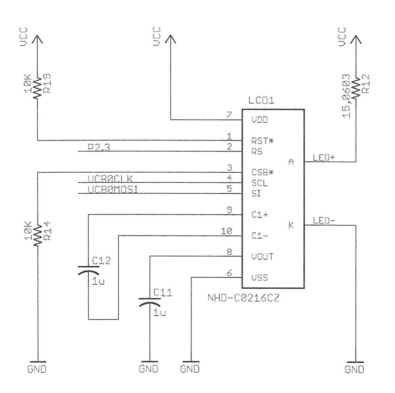

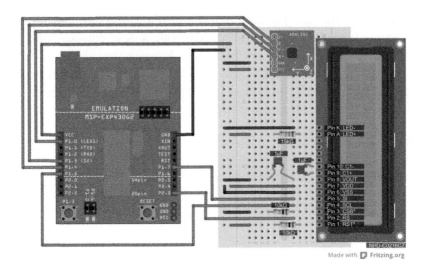

Made with ◾ Fritzing.org

Once your hardware is set up, let's write some code!

●●●

Tip

Or copy some code. This is a long piece of code, so if you don't feel like retyping 3 pages worth of code, you can go to the http://booksite.elsevier. com/9780124115880 and download the sketch in its entirety. We have placed the code in the book here for your reading convenience.

```
#include <SPI.h>
#include <LiquidCrystal.h>

const int xpin = A0;         // x-axis of the accelerometer
const int ypin = A3;         // y-axis
const int zpin = A4;         // z-axis
const int LCD_SS = 0;        // SS line is not used in our
                             // hardware setup
const int LCD_RS = P2_3;     // RS pin of our LCD
LiquidCrystal lcd(LCD_SS, LCD_RS);      // declare a LiquidCrystal
                                        // variable called "lcd"

String response = "";     // this will hold our magic LaunchPad response
int result = 0;           // this will help us randomly choose a response
int startTime;            // this will help us periodically read our
                          // accel data
void setup() {
    lcd.begin(16, 2);         // initialize our LCD
    lcd.print("Shake the magic");// print out intro text
    lcd.setCursor(0,1);       // reposition cursor to next row
    lcd.print("LaunchPad! ");

    startTime = millis();
    /* store start time. The millis() function returns how many
    milliseconds have elapsed since the program executed.*/
}
```

```
void loop() {
    /* these variables will hold the 3 axis accelerometer data sum
    will store our resultant vector */
    int a, b, c, sum;

    /* this checks to see if 100 ms have elapsed since we last checked
    accelerometer readings */
    if (millis()-startTime > 100){
        a = analogRead(xpin) - 512;   // read accel data
        b = analogRead(ypin) - 512;   // offset the data to "normalize"
                                      // each vector
        c = analogRead(zpin) - 512;

    /* use this formula to find the net force applied on all three
    vectors */
    sum = round(sqrt (a*a + b*b + c*c));
    if(sum>130) {    // check to see if we exceeded threshold
        /* if so, trigger the Magic LaunchPad to return a random
        message!*/
        result = random(0,3);
        /*random(min, max); This function returns a random value from
        [min, max-1]. Then, pass in the random number between 0 & 3
        into a switch statement to return random response */
        switch (result){
            case 0:
                response="Yes!";
                break;
            case 1:
                response="Maybe!";
                break;
            case 2:
                response="Nope!";
                break;
        }
        lcd.clear();// clear display to show new random response!
        lcd.setCursor(0,0);  // reset cursor
        lcd.print(response); // print our random response!
        delay(3000);     //show response for 3 seconds

        /*reset display and re-print the original prompt.*/
        lcd.setCursor(0,0);
        lcd.print("Shake the magic ");
        lcd.setCursor(0,1);
        lcd.print("LaunchPad!       ");
    }
    /* reset startTime. Wait for next 100ms before checking acc again.*/
    startTime=millis();
    }
}
```

```
void displayPrompt(){
    lcd.clear();// clear display to show new random response!
    lcd.setCursor(0,0);   // reset cursor
    response="Shake the magic ";
    lcd.print(response); // print our random response!
    response="LaunchPad!";
    lcd.setCursor(0,1);   // reposition cursor to next row
    lcd.print(response);
}
```

Once you have the code in your LaunchPad, compile, download and run the code! Then, think of a question for our LaunchPad to answer and give it a shake!

Challenges:

- Using the `millis()` function and your new SPI skills, can you create a stopwatch using the LaunchPad? Try using the pushbuttons on the LaunchPad itself to function as the start and stop button of your stopwatch.
- Use the potentiometer as a dial to create a tea timer, while the LCD module displays the amount of time left until the buzzer will alarm. Based on the position of the potentiometer, set the amount of time your LaunchPad should countdown from. Once the potentiometer is positioned for the right time, use the onboard pushbutton on the LaunchPad as a button to start and pause the countdown. Once the timer has elapsed and counted down to zero, buzz the buzzer!

10.13 OTHER LANGUAGES

We've spent quite a lot of time going through two types of serial communication protocols that our LaunchPad is capable of—UART and SPI. Not too shabby! But wait, there are even more languages in our LaunchPad's repertoire. One serial communication protocol that we won't go into much detail in this book is I2C, or Inter-Integrated Circuit. This is a serial communication module that is similar to SPI in that it is a synchronous interface that requires a shared clock signal between all communicating parties. One major difference of I2C, however, is that it is only a two-wire protocol, unlike SPI which is primarily a four-wire interface.

Despite having just two wires, I2C is capable of enabling multislave communication with a Master, much like SPI. One main difference, however, is that SPI was able to individually communicate with each slave using individual Slave Select lines. I2C accomplishes this in a different way. Instead, our LaunchPad can individually communicate with each slave through I2C by sending out a slave "address," or name before the actual command or payload. In addition to addressing each slave, the Master in the conversation can also specify if it wants to write (or talk) to a slave device or read from the slave device (aka listen).

I2C is capable of bi-directional communication and includes just two wires. One wire is to drive the clock signal (SCL) and the second wire is the data line (SDA). Once again, the Master drives the communication in this protocol.

10.14 WRAPPING UP THE LANGUAGES OF LAUNCHPAD

Our LaunchPad is a sociable device, capable of talking to external components such as LCDs, sensors, or other LaunchPads through various languages or serial communication protocols. We covered UART, SPI, and introduced I2C. We also learned how libraries can help us abstract away from the lower level communication protocols. This enables developers to focus more on the function and actions of our LaunchPad projects.

CHAPTER 11

Launchpad is Just the Beginning

11.1 WE DID IT

Congratulations, we have come a long way since we started our LaunchPad journey! Let's take a moment to look back and reflect on all the new concepts that we have discovered. We explored digital and analog signals, learned how our LaunchPad can make decisions based on conditionals, found out that our LaunchPad is multilingual, and more.

These valuable ideas will serve as fundamental building blocks for your future LaunchPad-based creations! In addition to introducing these concepts, the projects should have equipped you with a practical and systematic approach to solving problems and creating new things. By first analyzing the problem, then breaking it down into smaller achievable pieces, we are able to identify which tools and concepts we should apply to get the job done.

Being able to look at the problem systematically, able to break down the complex system to solvable pieces, isolate, and identify the components required to tackle each piece of the puzzle helps you understand how to leverage different features of the LaunchPad to solve the problem efficiently.

Just to give you a head start, the following are some of the common questions that can help you start with any application:

• What are the input signals for your system? Any digital buttons or switches? Do analog signals need to be measured, such as an analog sensor or a voltage from a potentiometer? If so, perhaps an integrated ADC might be needed.
• What are the output signals? Perhaps you need analog output? Are your digital outputs time-sensitive? How about a PWM signal to drive a piezo speaker, adjust an LED brightness, or control a motor?
• How do users interact with your product? Can the user interface (UI) be as simple as LEDs or does it require displays, or the help of a computer to relay this information?

- Do you need to communicate with other devices? Identify these components, and their preferred communication protocols to figure out the right system to fit your application.

Answering these questions can help you identify the resources you need, where to go to get more information, where and who to ask for more help, what kinds of testing/experiments you need to perform to figure out the solutions.

11.2 LOOKING BACK

At the beginning of the book we asked you to write down your dream application. Ask yourself the questions above and write down the answers. We want you to compare the answers with your original answers at the beginning of the book. Flip back to the beginning if you need a quick reminder. How much of it has changed? Are you able to now answer questions you previously could not with the new concepts and topics we covered in this book?

If the answer is yes, or even just a tiny bit, then you are heading in the right direction. To aid you on your way to create your ultimate LaunchPad application, here are a few more friendly places for information, knowledge, resources, and even some helping hands:

- LaunchPad homepage @ www.ti.com/launchpad
- Energia homepage for more examples, documentation, and installation @ www.energia.nu
- Home and wiki for the Educational BoosterPack @ www.boosterpackdepot.com
- The most helpful and friendly LaunchPad/MSP430 community on the internet @ www.43oh.com
- Checkout the complete list of BoosterPacks @ www.ti.com/boosterpacks
- Don't forget to use your favorite search engine! There are hundreds of thousands of LaunchPads out there, so be sure to use the world wide web if you encounter any questions.

With your new knowledge, we hope that this is just the beginning for something even greater. With a firm foundation in LaunchPad development, we hope that you are ready, prepared, and inspired to open up the next chapter of your microcontroller journey. Keep on learning and keep on making with your trusty LaunchPad!

Quick References

ENERGIA PIN MAPPING

The diagram below will prove to be a nifty "map" for your LaunchPad journey. There are few things to notice.

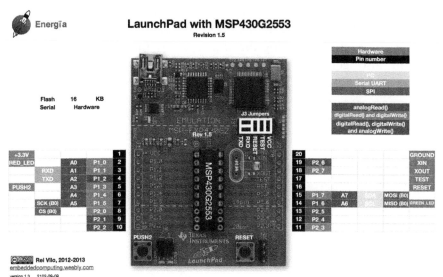

Pin numbering

The pins of the MSP430G2553 device that serves as our LaunchPad's "brain" has 20 pins. Each pin is numbered from 1 to 20, starting with pin 1 at the top left, and counts up counter-clockwise, ending with pin 20 at the top right.

Each pin has a unique set of functionality. Some can be used to read analog signals, while others can be used to generate analog signals. Some pins can also be used to enable serial communication with other components.

We can address each pin in Energia with some of the "nicknames" defined in the diagram. For example, we can address pin number 2 as:

- P1_0
- A0

- RED_LED
- 2

Available functionality at each pin

As mentioned, each pin on the LaunchPad has a varying set of capabilities. The map below helps us understand what functionality is available at each pin.

digitalRead() & digitalWrite()

The pins capable of reading or writing a HIGH or LOW signal are the pins highlighted in green or purple. We can read the presence of a voltage at an input pin using the `digitalRead()` function. Alternatively, we can also generate a HIGH or LOW digital signal with the `digitalWrite()` function. The functions are defined below.

```
digitalRead(pinNum);
/*
pinNum = the pin where you want to read a digital signal (P1_0, P1_1, P1_2,
P1_3, P1_4, P1_5, P1_6, P1_7, P2_0, P2_1, P2_2, P2_3, P2_4, P2_5, P2_6,
P2_7, or any associated "nickname")
*/
digitalWrite(pinNum, state);
/*
pinNum = the pin where you want to generate a digital signal (P1_0, P1_1,
P1_2, P1_3, P1_4, P1_5, P1_6, P1_7, P2_0, P2_1, P2_2, P2_3, P2_4, P2_5,
P2_6, P2_7, or any associated "nickname")
state = can be HIGH (1) or LOW (0), depending on the type of signal you
want to generate
*/
```

analogRead()

The pins capable of reading an analog signal using the `analogRead()` function in Energia are highlighted in blue. We can read an `analogRead()` function can be used as defined below:

```
analogRead(pinNum);
/*
pinNum = the pin where you want to read an analog signal
pinNum can be A0, A1, A2, A3, A4, A5, A6 or A7 (or any associated "nick-
name")
*/
```

```
analogWrite()
```

The pins capable of generating a PWM signal or an "average voltage" are highlighted in purple. The `analogWrite()` function accepts two parameters as shown below.

```
analogWrite(pinNum, intensity);
/*
pinNum 5 the pin where you want to generate a PWM/average voltage signal
pinNum can be P1_2, P1_6, P2_1, P2_2, P2_4, P2_5, P2_6 (or any associated
"nickname")
intensity5value within the range of [0, 255] to change the desired duty
cycle
*/
```

SPI-related pins

Also note the pins highlighted in red. These are the four pins used by SPI, which is typically a four-wire protocol. We can see the pins associated to SCK (the synchronous clock signal), CS (chip/slave select), MOSI (Master-Out/Slave-In signal), and MISO (Master-In/Slave-Out).

Remember that if you are using SPI to communicate to other modules like LCDs, sensors, etc., the other functionality available at the SPI pins may become unavailable.

UART-related pins

The pins shown in orange are associated to UART. Remember, this is the protocol that Energia uses for enabling simple debugging, allowing developers to send diagnostic data back to the computer, which can be displayed in the Serial Monitor. UART is typically a two-wire protocol enabling Full-Duplex/bidirectional communication. The RXD line and TXD line enable this communication and are in the "point-of-view" of our LaunchPad. So, the RXD line is the connection for computer-to-LaunchPad data, while TXD line is the connection for LaunchPad-to-computer data.

Remember that if you are using UART for debugging or other purposes, the other functionality available at that pin may become unavailable. If you are using UART for communicating with the computer, be sure that the RXD and TXD jumpers are connected appropriately on the LaunchPad at the header bank labelled "J4". For

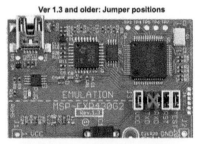

LaunchPad rev 1.4 or newer, populate the jumpers horizontally as shown in the image below. If you are using an older revision of the LaunchPad, remove the jumpers RXD and TXD and cross them using jumper cables.

USING THE EDUCATIONAL BOOSTERPACK

Here are some useful tips, tricks, and reminders when using the Educational BoosterPack.

Plugging in the Educational BoosterPack requires a few simple steps

1. Remove the jumpers on the LaunchPad that connect the green and red LEDs to the MSP430G2553 device. These jumpers are labelled as J5 and connect the LEDs to pin P1.0 (red) and P1.6 (green).
2. If you want to enable simple debugging over UART, be sure to position the RXD and TXD jumpers in the "HW UART" position (horizontal) on your LaunchPad. These are the two left-most jumpers and are part of the jumper set J3.
3. Now you can plug the Educational BoosterPack onto the LaunchPad!

Jumpers on the Educational BoosterPack

1. There are two jumpers on the Educational BoosterPack. These jumpers are there to enable the user to choose which analog signals they want to connect to the MSP430G2553 device that sits on the LaunchPad underneath.

   a. Jumper J5: Enables users to select between the Potentiometer and the Y-channel of the three-axis accelerometer to connect to pin A3 on the LaunchPad below.

   b. Jumper J6: Enables users to select between the Microphone and the Z-channel of the three-axis accelerometer to connect to pin A4 on the LaunchPad below.

   The X-axis channel of the accelerometer is always routed down to the LaunchPad and is connected to pin A0 on the LaunchPad.

Quick pin reference for the Educational BoosterPack

Here is a quick reference for how the LaunchPad pins are assigned to the Educational BoosterPack components.

- **A0** (connected to XOUT of accelerometer)
- **A3** (connected to YOUT of ACC OR output of potentiometer)
- **A4** (connected to ZOUT of ACC OR output of microphone)
- **UCB0CLK** (connected to the character LCD display)
- **UCB0SOMI** (connected to the character LCD display)
- **P2.1** (connected to the red channel of the RGB LED)
- **P2.2** (connected to the green channel of the RGB LED)
- **P2.3** (connected to the character LCD display "RS")
- **P2.4** (connected to the blue channel of the RGB LED)
- **P2.5** (connected to the white LED)
- **P2.6** (connected to the buzzer)
- **P2.7** (connected to alligator clip connector "MK1")

Finding more information for the Educational BoosterPack

You can learn more about the Educational BoosterPack including the pin assignment list above, schematics, PCB design files, and more at the Educational BoosterPack wiki: http://boosterpackdepot.com/wiki/index.php?title = Educational_BoosterPack.

LaunchPad with MSP430G2553

Revision 1.5

Energīa

Hardware		
Pin number		
I²C		
Serial UART		
SPI		
analogRead()		
digitalRead() and digitalWrite()		
digitalRead(), digitalWrite() and analogWrite()		

				GROUND
20				
19	P2_6			XIN
18	P2_7			XOUT
17				TEST
16				RESET
15	P1_7	A7	SDA	MOSI (B0)
14	P1_6	A6	SCL	MISO (B0) GREEN_LED
13	P2_5			
12	P2_4			
11	P2_3			

MSP430G2553

+3.3V					1
RED_LED	A0			P1_0	1
	A1	RXD		P1_1	2
	A2	TXD		P1_2	3
	A3			P1_3	4
	A4			P1_4	5
	A5	SCK (B0)		P1_5	6
		CS (B0)		P2_0	7
PUSH2				P2_1	8
				P2_2	9
					10

Flash   16   KB

Serial   Hardware

version 1.3   2102-09-09

Printed and bound by CPI Group (UK) Ltd, Croydon, CR0 4YY

03/10/2024

01040421-0017